Charles Francis Adams

Railroads

Their Origin And Problems

Charles Francis Adams

Railroads

Their Origin And Problems

ISBN/EAN: 9783744696395

Printed in Europe, USA, Canada, Australia, Japan

Cover: Foto ©berggeist007 / pixelio.de

More available books at **www.hansebooks.com**

RAILROADS:

THEIR ORIGIN AND PROBLEMS.

BY
CHARLES FRANCIS ADAMS, JR.

NEW YORK
G. P. PUTNAM'S SONS
182 FIFTH AVENUE
1878.

CONTENTS.

	PAGE
THE GENESIS OF THE RAILROAD SYSTEM................................	1
THE RAILROAD PROBLEM................................	80

THE GENESIS

OF THE

RAILROAD SYSTEM.

THERE are not many stories that are either more interesting in themselves or better worth telling for the lesson they convey, than the story of George Stephenson and his invention of the locomotive engine. It has been told, too, in a manner which upon the whole leaves little to be desired; and the great and long continued popularity of Smiles' biography is one of the most encouraging symptoms of the better and healthier education of the times. In the course of his narrative the author describes with great literary skill the genesis of the locomotive. In doing so he carries his readers along with him through episodes of opposition, discouragement, disappointment, almost defeat,—the interest in the narrative and the fortunes of its hero continually growing until it exceeds that of any work of fiction of the day, even though Walter Scott himself was then a living author,—until at last the great dramatic climax is reached in the memorable pageant of September 15th, 1830. That day,—the day of the formal opening of the Manchester & Liverpool railroad,—was for Stephenson more than an ovation, it was literally a triumph. Guiding his locomotive, the *Northumbrian*, at the head of the train, not only was he, even though the Duke of Wellington himself was there, the conquering hero observed of all, but there were also many circumstances about the occasion suggestive of other and less attractive features of the classic triumphs. Reminders of public distress and private want, of the fickleness of popular favor and of sudden death itself, all were there. The season was favorable, the skies were clear, the occasion great; but things would not move smoothly. It was a day of *contre-temps;* a day to be remembered and described, but one which nevertheless must ever after have left a bitter taste in the mental mouths of those who took part in its observances. Unfortunately, when he came to giving an account of it, Smiles' appreciation of the dramatic fitness of things proved too strong for his fidelity to facts. He thought his hero deserved a day of triumph unalloyed, and so he gave it to him—as nearly as he could. The terrible episode of Mr. Huskisson's death it was not possible to wholly pass over; but whatever else was there to mar the pleasure of the day could be ignored, and was ignored accordingly. The liberty with facts which Smiles thus allowed himself to take, was long since pointed out by Jeaffreson, in his excellent life of

Robert Stephenson; and in that work will be found a much more correct account than is given by Smiles of the events of the Manchester & Liverpool opening. Even Jeaffreson's account is, however, not wholly satisfactory. It was written too long after the event. He sees what he undertakes to describe with eyes accustomed to railroads and locomotives and trains of cars. He has with great industry gotten all his details together and woven them into a skilful narrative, but it is, after all, not the narrative of one who himself was there. Now the great peculiarity of the locomotive engine and its sequence, the railroad, as compared with other and far more important inventions, was that it burst rather than stole or crept upon the world. Its advent was in the highest degree dramatic. It was even more so than the discovery of America. Of this last we know every detail, and nothing is wanting which could lend an interest to the event. Picturesque and absorbing as the story is, however, the climax did not work itself out before the very eyes of an astonished world. Columbus and his crew alone on the morning of the 12th October 1492, saw the shores of the new world. And yet, next to the locomotive engine, this was probably the most dramatic of all those discoveries which have marked epochs in human history. The mariner's compass, far more momentous in its consequences, crept silently on a world which to this day does not know when or from whence or how it came. It was much the same with gun-powder. In the case of printing it is somewhat different, for though its invention has been a fruitful source of controversy, something at least is known of it. Hallam, indeed, in his Literature of Europe, indulges in a flight of rhetoric somewhat unusual with him and which reads queerly, as he speaks of Fustenburg's Mazarin Bible, the first printed book. "It is," he says, "a very striking circumstance, that the high-minded inventors of this great art tried at the very outset so bold a flight as the printing of an entire Bible, and executed it with astonishing success. It was Minerva leaping on earth in her divine strength and radiant armor, ready at the moment of her nativity to subdue and destroy her enemies.... We may see in imagination this venerable and splendid volume leading up the crowded myriads of its followers, and imploring, as it were, a blessing on the new art by dedicating its first fruits to the service of Heaven." "In imagination" perhaps, "we may see" all this, but assuredly the cotemporary world neither saw nor dreamed of it; on the contrary, imaginary processions apart, few things less inspiring can be conceived than the unnoticed homely toil of those poor German mechanics at Mentz, who four centuries ago launched upon an unconscious world the great motive power of all modern life. So with the loom, the steam engine, and electricity. Each and all, they struggled into existence slowly and painfully. The world never stopped to look, much less to wonder at them. We cannot know what people's sensations were when they first realized that a new power had appeared, for there was no particular moment at which they ever realized it. The locomotive engine alone as soon as it was seen was acknowledged; for it must be remembered that its one essential feature —the multitubular boiler,—was first used in Stephenson's experimental locomotive, the *Rocket*, on the Rainhill trial course in October, 1829, and never after that time was the importance of the new discovery denied, while the interest felt in its further development each day widened and became more engrossing.

It was this element of spontaneity, therefore,— the instantaneous and dramatic recognition of success, which gave a peculiar interest to everything connected with the Manchester & Liverpool railroad. The whole world was looking at it, with a full realizing sense that something great and momentous was impending. Every day people watched the gradual development of the thing, and actually took part in it. In doing so they had sensations and those sensations they have described. There is consequently an element of human nature surrounding it. The complete ignoring of this element by both Smiles and Jeaffreson is a defect in their narratives. They describe the scene from a standpoint of forty years later. Others described it as they saw it at the time. To their descriptions time has only lent a new freshness. They are full of honest wonder. They are much better and more valuable and more interesting now than they were fifty years ago, and for that reason are well worth exhuming.

To introduce the contemporaneous story of the day, however, it is not necessary even to briefly review the long series of events which had slowly led up to it. The world is tolerably familiar with the early life of George Stephenson, and with the vexatious obstacles he had to overcome before he could even secure a trial for his invention. The man himself, however, is an object of a good deal more curiosity to us, than he was to those among whom he lived and moved. A living glimpse at him now is worth dwelling upon, and is the best possible preface to any account of his great day of life triumph. Just such a glimpse of the man has been given to us at the moment when at last all difficulties had been overcome,—when the Manchester & Liverpool railroad was completed; and, literally, not only the eyes of Great Britain but those of all civilized countries were directed to it and to him who had originated it. At just that time it chanced that the celebrated actor, John Kemble, was fulfilling an engagement at Liverpool with his daughter, better known in this country as Mrs. Frances Kemble Butler. The extraordinary social advantages the Kemble family enjoyed gave both father and daughter opportunities such as seldom came in the way of ordinary mortals. For the time being they were, in fact, the lions of the stage, just as George Stephenson was the lion of the new railroad. As was most natural the three lions were brought together. The young actress has since published her impressions, jotted down at the time, of the old engineer. Her account of a ride side by side with George Stephenson, on the seat of his locomotive, over the as yet unopened road, is one of the most interesting and life-like records we have of the man and the enterprise. Perhaps it is the most interesting. The introduction is Mrs. Kemble's own, and written forty-six years after the experience:

"While we were acting at Liverpool, an experimental trip was proposed upon the line of railway which was being constructed between Liverpool and Manchester, the first mesh of that amazing iron net which now covers the whole surface of England, and all civilized portions of the earth. The Liverpool merchants, whose far-sighted self-interest prompted to wise liberality, had accepted the risk of George Stephenson's magnificent experiment, which the committee of inquiry of the House of Commons had rejected for the government. These men, of less intellectual culture than the Parliament members, had the adventurous imagination proper to great speculators, which is the poetry of the counting house and wharf, and were better able to receive the enthusiastic infection of the great projector's sanguine hope than the Westminster committee. They were exultant and triumphant at the near completion of the work, though, of course not without some misgivings as to the eventual success of the stupendous enterprise. My father knew several of the gentlemen most deeply interested in the undertaking, and Stephenson having proposed a trial trip as far as the fifteen-mile viaduct, they, with infinite kindness, invited him and permitted me to accompany them: allowing me, moreover, the place which I felt to be one of supreme honor, by the side of Stephenson. All that wonderful history, as much more interesting than a romance as truth is stranger than fiction, which Mr. Smiles' biography of the projector has given in so attractive a form to the world, I then heard from his own lips. He was rather a stern-featured man, with a dark and deeply marked countenance; his speech was strongly inflected with his native Northumbrian accent, but the fascination of that story told by himself, while his tame dragon flew panting along his iron pathway with us, passed the first reading of the Arabian Nights, the incidents of which it almost seemed to recall. He was wonderfully condescending and kind, in answering all the questions of my eager ignorance, and I listened to him with eyes brimful of warm tears of sympathy and enthusiasm, as he told me of all his alternations of hope and fear, of his many trials and disappointments, related with fine scorn, how the "Parliament men" had badgered and baffled him with their book-knowledge, and how, when at last they had smothered the irrepressible prophecy of his genius in the quaking depths of Chat Moss, he had exclaimed, 'Did ye ever see a boat float on water? I will make my road float upon Chat Moss!' The well-read Parliament men (some of whom, perhaps, wished for no railways near their parks and pleasure-grounds) could not believe the miracle, but the shrewd Liverpool merchants, helped to their faith by a great vision of immense gain, did; and so the railroad was made, and I took this memorable ride by the side of its maker, and would not have exchanged the honor and pleasure of it for one of the shares in the speculation."

"LIVERPOOL, August 26th, 1830.

"MY DEAR H——: A common sheet of paper is enough for love, but a foolscap extra can only contain a railroad and my ecstasies. There was once a man born at Newcastle-upon-Tyne, who was a common coal-digger; this man had an immense constructiveness, which displayed itself in pulling his watch to pieces and putting it together again; in making a pair of shoes when he happened to be some days without occupation;

finally—here there is a great gap in my story—it brought him in the capacity of an engineer before a committee of the House of Commons, with his head full of plans for constructing a railroad from Liverpool to Manchester. It so happened that to the quickest and most powerful perceptions and conceptions, to the most indefatigable industry and perseverance, and the most accurate knowledge of the phenomena of nature as they affect his peculiar labors, this man joined an utter want of the 'gift of gab;' he could no more explain to others what he meant to do and how he meant to do it, than he could fly, and therefore the members of the House of Commons, after saying, 'There is a rock to be excavated to a depth of more than sixty feet, there are embankments to be made nearly to the same height, there is a swamp of five miles in length to be traversed, in which if you drop an iron rod it sinks and disappears; how will you do all this?' and receiving no answer but a broad Northumbrian, 'I can't tell you how I'll do it, but I can tell you I *will* do it,' dismissed Stephenson as a visionary. Having prevailed upon a company of Liverpool gentlemen to be less incredulous, and having raised funds for his great undertaking, in December of 1826 the first spade was struck into the ground. And now I will give you an account of my yesterday's excursion. A party of sixteen persons was ushered into a large court-yard, where, under cover, stood several carriages of a peculiar construction, one of which was prepared for our reception. It was a longbodied vehicle with seats placed across it back to back; the one we were in had six of these benches, and was a sort of uncovered *char à banc*. The wheels were placed upon two iron bands, which formed the road, and to which they are fitted, being so constructed as to slide along without any danger of hitching or becoming displaced, on the same principle as a thing sliding on a concave groove. The carriage was set in motion by a mere push, and, having received this impetus, rolled with us down an inclined plane into a tunnel, which forms the entrance to the railroad. This tunnel is four hundred yards long (I believe), and will be lighted by gas. At the end of it we emerged from darkness, and, the ground becoming level, we stopped. There is another tunnel parallel with this, only much wider and longer, for it extends from the place we had now reached, and where the steam carriages start, and which is quite out of Liverpool, the whole way under the town, to the docks. This tunnel is for wagons and other heavy carriages; and as the engines which are to draw the trains along the railroad do not enter these tunnels, there is a large building at this entrance which is to be inhabited by steam engines of a stationary turn of mind, and different constitution from the traveling ones, which are to propel the trains through the tunnels to the terminus in the town, without going out of their houses themselves. The length of the tunnel parallel to the one we passed through is (I believe) two thousand two hundred yards. I wonder if you are understanding one word I am saying all this while! We were introduced to the little engine which was to drag us along the rails. She (for they make these curious little fire horses all mares) consisted of a boiler, a stove, a platform, a bench, and behind the bench a barrel containing enough water to prevent her being thirsty for fifteen miles,—the whole machine not bigger than a common fire engine. She goes upon two wheels, which are her feet, and are moved by bright steel legs called pistons; these are propelled by steam, and in proportion as more steam is applied to the upper extremities (the hip-joints, I suppose) of these pistons, the faster they move the wheels; and when it is desirable to diminish the speed, the steam, which unless suffered to escape would burst the boiler, evaporates through a safety valve into the air. The reins, bit, and bridle of this wonderful beast, is a small steel handle, which applies or withdraws the steam from its legs or pistons, so that a child might manage it.

"The coals, which are its oats, were under the bench, and there was a small glass tube affixed to the boiler, with water in it, which indicates by its fullness or emptiness when the creature wants water, which is immediately conveyed to it from its reservoirs. There is a chimney to the stove, but as they burn coke there is none of the dreadful black smoke which accompanies the progress of a steam vessel. This snorting little animal, which I felt rather inclined to pat, was then harnessed to our carriage, and Mr. Stephenson having taken me on the bench of the engine with him, we started at about ten miles an hour. The steam horse being ill adapted for going up and down hill, the road was kept at a certain level, and appeared sometimes to sink below the surface of the earth and sometimes to rise above it. Almost at starting it was cut through the solid rock, which formed a wall on either side of it, about sixty feet high. You can't imagine how strange it seemed to be journeying on thus, without any visible cause of progress other than the magical machine, with its flying white breath and rhythmical, unvarying pace, between these rocky walls, which are already clothed with moss and ferns and grasses; and when I reflected that these great masses of stone had been cut asunder to allow our passage thus far below the surface of the earth, I felt as if no fairy tale was ever half so wonderful as what I saw. Bridges were thrown from side to side across the top of these cliffs, and the people looking down upon us from them seemed like pigmies standing in the sky. I must be more concise, though, or I shall want room. We were to go only fifteen miles, that distance being sufficient to show the speed of the engine, and to take us to the most beautiful and wonderful object on the road. After proceeding through this rocky defile, we presently found ourselves raised upon embankments ten or twelve feet high; we then came to a moss, or swamp, of considerable extent, on which no human foot could tread without sinking, and yet it bore the road which bore us. This had been the great stumbling-block in the minds of the committee of the House of Commons; but Mr. Stephenson has succeeded in overcoming it. A foundation of hurdles, or, as he called it, basket-work, was thrown over the morass, and the interstices were filled with moss and other elastic matter.

Upon this the clay and soil were laid down, and the road does float, for we passed over it at the rate of five and twenty miles an hour, and saw the stagnant swamp water trembling on the surface of the soil on either side of us. I hope you understand me. The embankment had gradually been rising higher and higher, and in one place, where the soil was not settled enough to form banks, Stephenson had constructed artificial ones of woodwork, over which the mounds of earth were heaped, for he said that though the wood-work would rot, before it did so the banks of earth which covered it would have been sufficiently consolidated to support the road. We had now come fifteen miles, and stopped where the road traversed a wide and deep valley. Stephenson made me alight and led me down to the bottom of this ravine, over which, in order to keep his road level, he has thrown a magnificent viaduct of nine arches, the middle one of which is seventy feet high, through which we saw the whole of this beautiful little valley. It was lovely and wonderful beyond all words. He here told me many curious things respecting this ravine; how he believed the Mersey had once rolled through it; how the soil had proved so unfavorable for the foundation of his bridge that it was built upon piles, which had been driven into the earth to an enormous depth; how while digging for a foundation he had come to a tree bedded in the earth, fourteen feet below the surface of the ground; how tides are caused, and how another flood might be caused; all of which I have remembered and noted down at much greater length than I can enter upon it here. He explained to me the whole construction of the steam-engine, and said he could soon make a famous engineer of me, which, considering the wonderful things he has achieved, I dare not say is impossible. His way of explaining himself is peculiar, but very striking, and I understood, without difficulty, all that he said to me. We then rejoined the rest of the party, and the engine having received its supply of water, the carriage was placed behind it, for it cannot turn, and was set off at its utmost speed, thirty-five miles an hour, swifter than a bird flies (for they tried the experiment with a snipe). You cannot conceive what that sensation of cutting the air was; the motion is as smooth as possible, too. I could either have read or written; and as it was, I stood up, and with my bonnet off 'drank the air before me.' The wind, which was strong, or perhaps the force of our own thrusting against it, absolutely weighed my eyelids down.

"When I closed my eyes this sensation of flying was quite delightful, and strange beyond description; yet strange as it was, I had a perfect sense of security, and not the slightest fear. At one time, to exhibit the power of the engine, having met another steam-carriage which was unsupplied with water, Mr. Stephenson caused it to be fastened in front of ours; moreover, a wagon laden with timber was also chained to us, and thus propelling the idle steam-engine, and dragging the loaded wagon which was beside it, and our own carriage full of people behind, this brave little she-dragon of ours flew on. Farther on she met three carts, which, being fastened in front of her, she pushed on before her without the slightest delay or difficulty; when I add that this pretty little creature can run with equal facility either backwards or forwards, I believe I have given you an account of all her capacities. Now for a word or two about the master of all these marvels, with whom I am most horribly in love. He is a man from fifty to fifty-five years of age; his face is fine, though careworn, and bears an expression of deep thoughtfulness; his mode of explaining his ideas is peculiar and very original, striking, and forcible; and although his accent indicates strongly his north country birth, his language has not the slightest touch of vulgarity or coarseness. He has certainly turned my head. Four years have sufficed to bring this great undertaking to an end. The railroad will be opened upon the fifteenth of next month. The Duke of Wellington is coming down to be present on the occasion, and, I suppose, what with the thousands of spectators and the novelty of the spectacle, there will never have been a scene of more striking interest. The whole cost of the work (including the engines and carriages) will have been eight hundred and thirty thousand pounds ; and it is already worth double that sum. The directors have kindly offered us three places for the opening, which is a great favor, for people are bidding almost anything for a place, I understand."

Even while Miss Kemble was writing this letter, certainly before it had reached her correspondent, the official programme of that opening to which she was so eagerly looking forward was thus referred to in the Liverpool papers:

"The day of opening still remains fixed for Wednesday the fifteenth instant. The company by whom the ceremony is to be performed, is expected to amount to eight or nine hundred persons, including the Duke of Wellington and several others of the nobility. They will leave Liverpool at an early hour in the forenoon, probably ten o'clock, in carriages drawn by eight or nine engines, including the new engine of Messrs. Braithwaite and Ericsson, if it be ready in time. The other engines will be those constructed by Mr. Stephenson, and each of them will draw about a hundred persons. On their arrival at Manchester, the company will enter the upper stories of the warehouses by means of a spacious outside wooden staircase, which is in course of erection for the purpose by Mr. Bellhouse. The upper story of the range of warehouses is divided into five apartments, each measuring sixty-six feet by fifty-six. In four of these a number of tables (which Mr. Bellhouse is also preparing) will be placed, and the company will partake of a splendid cold collation which is to be provided by Mr. Lynn, of the Waterloo Hotel, Liverpool. A large apartment at the east end of the warehouses will be reserved as a withdrawing room for the ladies, and is partitioned off for that purpose. After partaking of the hospitality of the directors, the company will return to Liverpool in the same order in which they arrive. We understand that each shareholder in the railway will be entitled to a seat (transferable) in one of the carriages, on this interesting and important occasion. It may be proper to state, for the information of the public, that no one will be permitted to go upon the railway between Ordsall lane and the warehouses, and parties of the military and police will be placed to preserve order, and prevent intrusion. Beyond Ordsall lane, however, the public will be freely admitted to view the procession as it passes: and no restriction will be laid upon them farther than may be requisite to prevent them from approaching too close to the rails, lest accidents should occur. By extending themselves along either side of the road towards Eccles any number of people, however great, may be easily accommodated."

It only remained to successfully carry out on the the 15th the programme thus carefully laid down. Of their ability so to do the directors of the company probably entertained little doubt. Yet there were circumstances connected with the then condition of public affairs which might well have occasioned them some uneasiness. Never in modern times had England passed through a sadder or more anxious period than that during which the Manchester & Liverpool road was built. The great reaction which naturally followed the close of the long Napoleonic wars was coming to a close, and the patience of all, and the endurance of many, were thoroughly worn out. The suffering of the poorer classes, especially in the manufacturing districts, was extremely severe, and the consequent popular discontent so great that even the semblance of order was with difficulty preserved. Half the counties in England were nightly illumined by incendiary fires. A fierce political agitation was also raging. The Duke of Wellington was prime minister. The cry for parliamentary reform was loud, and against any compliance with that cry the prime minister had set his face like a flint. From being the most popular man in the kingdom, he had become the most unpopular. He lived in constant danger of being hustled wherever he showed himself, even if he escaped mobbing. And now this man, hard, ungracious in manner, unyielding as iron, the object of intense popular odium, was coming down into the very hot-bed of suffering and agitation to take the prominent part,—to be the guest of honor upon an occasion which was sure to call out the entire mass of the population. Whether the directors of the company realized it or no, the experiment was a perilous one. In spite of every precaution the day might not improbably end in a riot,—possibly in a revolution. At last it came, and the contemporaneous reporter has left of it the following account:

"The town itself [Liverpool] was never so full of strangers; they poured in during the last and the beginning of the present week from almost all parts of the three kingdoms, and we believe that through Chester alone, which is by no means a principal road to Liverpool, four hundred extra passengers were forwarded on Tuesday. All the inns in the town were crowded to overflowing, and carriages stood in the streets at night, for want of room in the stable yards.

"On the morning of Wednesday the population of the town and of the country began very early to assemble near the railway. The weather was favorable, and the Company's station at the boundary of the town was the rendezvous of the nobility and gentry who attended, to form the procession at Manchester. Never was there such an assemblage of rank, wealth, beauty, and fashion in this neighborhood. From before nine o'clock until ten the entrance in Crown street was thronged by the splendid equipages from which the company was alighting, and the area in which the railway carriages were placed was gradually filling with gay groups eagerly searching for their respective places, as indicated by numbers corresponding with those on their tickets. The large and elegant car constructed for the nobility, and the accompanying cars for the Directors and the musicians were seen through the lesser tunnel, where persons moving about at the far end appeared as diminutive as if viewed through a concave glass. The effect was singular and striking. In a short time all those cars were brought along the tunnel into the yard which then contained all the carriages, which were to be attached to the eight locomotive engines which were in readiness beyond the tunnel in the great excavation at Edge-hill. By this time the area presented a beautiful spectacle, thirty-three carriages being filled by elegantly dressed persons, each train of carriages being distinguished by silk flags of different colors; the band of the fourth King's Own Regiment, stationed in the adjoining area, playing military airs, the Wellington Harmonic Band, in a Grecian car for the procession, performing many beautiful miscellaneous pieces; and a third band occupying a stage above Mr. Harding's Grand Stand, at William the Fourth's Hotel, spiritedly adding to the liveliness of the hour whenever the other bands ceased.

"A few minutes before ten, the discharge of a gun and the cheers of the assembly announced the arrival of the Duke of Wellington, who entered the area with the Marquis and Marchioness of Salisbury and a number of friends, the band playing "See the conquering Hero comes." He returned the congratulations of the company, and in a few moments the grand car, which he and the nobility and the principal gentry occupied, and the cars attached to it, were permitted to proceed; we say permitted, because no applied power, except a slight impulse at first is requisite to propel carriages along the tunnel, the slope being just sufficient to call into effect the principle of gravitation. The tunnel was lighted with gas, and the motion in passing through it must have been as pleasing as it was novel to all the party. On arriving at the engine station, the cars were attached to the *Northumbrian*, locomotive engine, on the southern of the two lines of rail; and immediately the other trains of carriages started through the tunnel and were attached to their respective engines on the northern of the lines.

"We had the good fortune to have a place in the first train after the grand cars, which train, drawn by the *Phœnix*, consisted of three open and two close carriages, each carrying twenty-six ladies and gentlemen. The lofty banks of the engine station were crowded with thousands of spectators, whose enthusiastic cheering seemed to rend the air. From this point to Wavertree-lane, while the procession was forming, the grand cars passed and repassed the other trains of carriages several times, running as they did in the same direction on the two parallel tracks, which gave the assembled thousands and tens of thousands the opportunity of seeing distinctly the illustrious strangers, whose presence gave extraordinary interest to the scene. Some soldiers of the 4th Regiment assisted the railway police in keeping the way clear and preserving order, and they discharged their duty in a very proper manner. A few minutes before eleven all was ready for the journey, and certainly a journey upon a railway is one of the most delightful that can be imagined. Our first thoughts it might be supposed, from the road being so level, were that it must be monotonous and uninteresting. It is precisely the contrary; for as the road does not rise and fall like the ground over which we pass, but proceeds nearly at a level, whether the land be high or low, we are at one moment drawn through a hill, and find ourselves seventy feet below the surface, in an Alpine chasm, and at another we are as many feet above the green fields, traversing a raised path, from which we look down upon the roofs of frame houses, and see the distant hills and woods. These variations give an interest to such a journey which cannot be appreciated until they are witnessed. The signal gun being fired, we started in beautiful style, amidst the deafening plaudits of the well-dressed people who thronged the numerous booths, and all the walls and eminences on both sides the line. Our speed was gradually increased till, entering the Olive Mountain excavation, we rushed into the awful chasm at the rate of twenty-four miles an hour. The banks, the bridges over our heads, and the rude projecting corners along the sides, were covered with masses of human beings past whom we glided as if upon the wings of the wind. We soon came into the open country of Broad Green, having fine views of Huyton and Prescot on the left, and the hilly grounds of Cheshire on the right. Vehicles of every description stood in the fields on both sides, and thousands of spectators still lined the margin of the road; some horses seemed alarmed, but after trotting with their carriages to the farther hedges, they stood still as if their fears had subsided. After passing Whiston, sometimes going slowly, sometimes swiftly, we observed that a vista formed by several bridges crossing the road gave a pleasing effect to the view. Under Rainhill Bridge, which, like all the others, was crowded with spectators, the Duke's car stopped until we passed, and on this, as on similar occasions, we had excellent opportunities of seeing the whole of the noble party, distinguishing the Marquis and Marchioness of Salisbury, the Earl and Countess of Wilton, Lord Stanley and others, in the fore part of the car; along side of the latter part was Mr. Huskisson, standing with his face always toward us; and further behind was Lord Hill, and others, among whom the Mayor of Liverpool took his station. At this place Mr. Bretherton had a large party of friends in a field, overlooking the road. As we approached the Sutton inclined plane the Duke's car passed us again at a most rapid rate —it appeared rapid even to us who were travelling then at, probably, fifteen miles an hour. We had a fine view of Billings hill from this neighborhood, and of a thousand various colored fields.

A grand stand was here erected, beautifully decorated, and crowded with ladies and gentlemen from St. Helen's and the neighborhood. Entering upon Parr Moss we had a good view of Newton Race Course and the stands, and at this time the Duke was far ahead of us; the grand cars appeared actually of diminutive dimensions, and in a short time we saw them gliding beautifully over the Sankey Viaduct, from which a scene truly magnificent lay before us.

"The fields below us were occupied by thousands who cheered us as we passed over the stupendous edifice; carriages filled the narrow lanes, and vessels in the water had been detained in order that their crews might gaze up at the gorgeous pageant passing far above their mast heads. Here again was a grand stand, and here again enthusiastic plaudits almost deafened us. Shortly, we passed the borough of Newton, crossing a fine bridge over the Warrington road, and reached Parkside, seventeen miles from Liverpool, in about four minutes under the hour. At this place the engines were ranged under different watering stations to receive fresh water, the whole extending along nearly a half a mile of road. Our train and two others passed the Duke's car, and we in the first train had had our engine supplied with water, and were ready to start, some time before we were aware of the melancholy cause of our apparently great delay. We had, most of us, alighted, and were walking about, congratulating each other generally, and the ladies particularly, on the truly delightful treat we were enjoying, all hearts bounding with joyous excitement, and every tongue eloquent in the praise of the gigantic work now completed, and the advantages and pleasures it afforded. A murmur and an agitation at a little distance betokened something alarming and we too soon learned the nature of that lamentable event, which we cannot record without the most agonized feelings. On inquiring, we learnt the dreadful particulars. After three of the engines with their trains had passed the Duke's carriage, although the others had to follow, the company began to alight from all the carriages which had arrived.

The Duke of Wellington and Mr. Huskisson had just shaken hands, and Mr. Huskisson, Prince Esterhazy, Mr. Birch, Mr. H. Earle, Mr. William Holmes, M. P. and others were standing in the road, when the other carriages were approaching. An alarm being given, most of the gentlemen sprang into the carriage, but Mr. Huskisson seemed flurried, and from some cause, not clearly ascertained, he fell under the engine of the approaching carriages, the wheel of which shattered his leg in the most dreadful manner. On being raised from the ground by the Earl of Wilton, Mr. Holmes, and other gentlemen, his only exclamations were; —" Where is Mrs. Huskisson? I have met my death. God forgive me." Immediately after he swooned. Dr. Brandreth, and Dr. Southey, of London, immediately applied bandages to the limb. In a short time the engine was detached from the Duke's carriage, and the musician's car being prepared for the purpose, the Right Honorable gentleman was placed in it, accompanied by his afflicted lady, with Doctor Brandreth, Dr. Southey, Earl of Wilton, and Mr. Stephenson, who set off in the direction of Manchester.

" The whole of the procession remained at least another hour uncertain what course to adopt. A consultation was held on the open part of the road, and the Duke of Wellington was soon surrounded by the Directors, and a mournful group of gentlemen. At first it was thought advisable to return to Liverpool, merely dispatching one engine and a set of carriages, to convey home Lady Wilton, and others who did not wish to return to Liverpool. The Duke of Wellington and Sir Robert Peel seemed to favor this course; others thought it best to proceed as originally intended: but no decision was made till the Boroughreeve of Manchester stated, that if the procession did not reach Manchester, where an unprecedented concourse of people would be assembled, and would wait for it, he should be fearful of the consequences to the peace of the town. This turned the scale and his Grace then proposed that the whole party should proceed, and return as soon as possible, all festivity at Manchester being avoided. The *Phœnix*, with its train, was then attached to the *North Star* and its train, and from the two united a long chain was affixed to his Grace's car, and although it was on the other line of rail, it was found to draw the whole along exceedingly well. About half-past one, we resumed our journey; and we should here mention that the Wigan Branch Railway Company had erected near Parkside bridge, a grand stand, which they and their friends occupied, and from which they enthusiastically cheered the procession. On reaching the twentieth mile post we had a beautiful view of Rivington Pike, and Blackstone Edge, and at the twenty-first the smoke of Manchester appeared to be directly at the termination of our view. Groups of people continued to cheer us, but we could not reply; our enjoyment was over. Tyldesley Church, and a vast region of smiling fields here met the eye, as we traversed the flat surface of Chat Moss, in the midst of which a vast crowd was assembled to greet us with their plaudits; and from the twenty-fourth mile post we began to find ourselves flanked on both sides by spectators extending in a continuous and thickening body all the way to Manchester. At the twenty-fifth mile post we met Mr. Stephenson returning with the Northumbrian engine. In answer to innumerable and eager inquiries, Mr. Stephenson said he had left Mr. Huskisson at the house of the Rev. Mr. Blackburn, Vicar of Eccles, and had then proceeded to Manchester, whence he brought back medical assistance, and that the surgeons, after seeing Mr. Huskisson, had expressed a hope that there was no danger. Mr. Stephenson's speed had been at the rate of thirty-four miles an hour during this painful errand. The engine being then again attached to the Duke's car, the procession dashed forward, passing countless thousands of people upon house tops, booths, high ground, bridges, etc., and our readers must imagine, for we cannot describe, such a movement through an avenue of living beings, and extending six miles in length. Upon one bridge a tri-colored flag was displayed; near another the motto of "Vote by ballot" was seen; in a field near Eccles, a poor and wretchedly-dressed man had his loom close to the roadside, and was weaving with all his might; cries of "No Corn Laws," were occasionally heard, and for about two miles the cheerings of the crowd were interspersed with a continual hissing and hooting from the minority. On approaching the bridge which crosses the Irwell, the 59th regiment was drawn up, flanking the road on each side, and presenting arms as his Grace passed along. We reached the warehouses at a quarter before three, and those who alighted were shown into the large upper rooms where a most elegant cold collation had been prepared by Mr. Lynn, for more than one thousand persons. The greater portion of the company, as the carriages continued to arrive, visited the rooms and partook in silence of some refreshment. They then returned to their carriages which had been properly placed for returning. His Grace and the principal party did not alight; but he went through a most fatiguing office for more than an hour and a half, in shaking hands with thousands of people, to whom he stooped over the hand rail of the carriage, and who seemed insatiable in their desire to join hands with him. Many women brought their children to him, lifting them up that he might bless them, which he did, and during the whole time he had scarcely a minute's respite. At half past four the Duke's car began to move away for Liverpool.

"They would have been detained a little longer, in order that three of the engines, which had been to Eccles for water, might have dropped into the rear to take their places; but Mr. Lavender represented that the crowd was so thickening in upon all sides, and becoming so clamorous for admission into the area, that he would not answer for the peace of the town, if further delay took place. The three engines were on the same line of rail as the Duke, and they could not cross to the other line without getting to a turning place, and as the Duke could not be delayed on account of his keeping the crowd together, there was no alternative but to send the engines forward. One of the other engines was then attached to our train, and we followed the Duke rapidly, while the six trains behind had only three engines left to bring them back. Of course, we kept pace with the Duke, who stopped at Eccles to inquire after Mr. Huskisson. The answer received was that there was now no hope of his life being saved; and this intelligence plunged the whole party into still deeper distress. We proceeded without meeting any fresh incident, until we passed Prescot, where we found two of the three engines at the 6¼ mile post, where a turning had been effected, but the third had gone on to Liverpool; we then detached the one we had borrowed, and the three set out to meet the six remaining trains of carriages. Our carriages were then connected with the grand cars, the engine of which now drew the whole number of nine carriages, containing nearly three hundred persons, at a very smart rate. We were now getting into vast crowds of people, most of them ignorant of the dreadful event which had taken place, and all of them giving us enthusiastic cheers which we could not return.

"At Roby, his Grace and the Childwalls alighted and proceeded home; our carriages then moved forward to Liverpool, where we arrived about seven o'clock, and went down the great tunnel, under the town, a part of the work which, more than any other, astonished the numerous strangers present. It is, indeed, a wonderful work, and makes an impression never to be effaced from the memory. The Company's yard, from Saint James's street to Wapping, was filled with carriages waiting for the returning parties, who separated with feelings of mingled gratification and distress, to which we shall not attempt to give utterance. We afterwards learnt that the parties we left at Manchester placed the three remaining engines together, and all the carriages together, so as to form one grand procession, including twenty-four carriages, and were coming home at a steady pace, when they were met near Newton by the other three engines, which were then attached to the rest, and they arrived in Liverpool about ten o'clock.

"Thus ended a pageant, which, for importance as to its object and grandeur in its details, is admitted to have exceeded anything ever witnessed. We conversed with many gentlemen of great experience in public life, who spoke of the scene as surpassing anything they had ever beheld, and who computed, upon data which they considered to be satisfactory, that not fewer than 500,000 persons must have been spectators of the procession."

So far from being a success, the occasion was, after the accident to Mr. Huskisson, such a series of mortifying disappointments and the Duke of Wellington's experience at Manchester had been so very far removed from gratifying, that the directors of the company felt moved to exonerate themselves from the load of censure by an official explanation. This they did in the following language :

" On the subject of delay which took place in the starting from Manchester, and consequently in the arrival at Liverpool, of the last three engines, with twenty-four carriages and six hundred passengers, being the train allotted to six of the engines, we are authorized to state that the directors think it due to the proprietors and others constituting the large assemblage of company in the above trains to make known the following particulars :

" Three out of the six locomotive engines, which belonged to the above teams, had proceeded on the south road from Manchester to Eccles, to take in water, with the intention of returning to Manchester, and so getting out of that line of road before any of the trains should start on their return home. Before this, however, was accomplished, the following circumstances seemed to render it imperative for the train of carriages, containing the Duke of Wellington and a great many of the distinguished visitors to leave Manchester. The eagerness on the part of the crowd to see the Duke, and to shake hands with him, was very great, so much so, that his Grace held out both his hands to the pressing multitude at the same time : the assembling crowd becoming more dense every minute, closely surrounded the carriages, as the principal attraction was this particular train. The difficulty of proceeding at all increased every moment and consequently the danger of accident upon the attempt being made to force a way through the throng also increased. At this juncture Mr. Lavender, the head of the police establishment of Manchester, interfered, and entreated that the Duke's train should move on, or he could not answer for the consequences. Under these circumstances, and the day being well advanced, it was thought expedient at all events to move forward while it was still practicable to do so. The order was accordingly given, and the train passed along out of the immediate neighborhood of Manchester without accident to any one. When they had proceeded a few miles they fell in with the engines belonging to the trains left at Manchester, and these engines being on the same line as the carriages of the procession, there was no alternative but bringing the Duke's train back through the dense multitude to Manchester, or proceeding with three extra engines to the neighborhood of Liverpool, (all passing places from one road to the other being removed, with a view to safety, on the occasion,) and afterwards sending them back to the assistance of the trains unfortunately left behind. It was determined to proceed towards Liverpool, as being decidedly the most advisable course under the circumstances of the case ; and it may be mentioned for the satisfaction of any party who may have considered that he was in some measure left in the lurch, that Mr. Moss, the Deputy Chairman, had left Mrs. Moss and several of his family to come with the trains which had been so left behind. Three engines having to draw a load calculated for six, their progress was of course much retarded, besides a considerable delay which took place before the starting of the last trains, owing to the uncertainty which existed as to what had become of the three missing engines. These engines, after proceeding to within a few miles of Liverpool, were enabled to return to

Parkside (in the neighborhood of Newton) where they were attached to the other three, and the whole proceeding safely to Liverpool, where they arrived at ten in the evening."

The case was, however, here stated, to say the least, in the mildest possible manner. The fact was that the authorities at Manchester had, and not without reason, passed a very panic-stricken hour on account of the Duke of Wellington. That personage had been in a position of no inconsiderable peril. Though the reporter preserved a decorous silence on that point, the ministerial car had on the way been pelted, as well as hooted; and at Manchester a vast mass of not particularly well disposed persons had fairly overwhelmed both police and soldiery, and had taken complete possession of the tracks. They were not riotous, but they were very rough; and they insisted on climbing upon the carriages and pressing their attentions on the distinguished inmates in a manner somewhat at variance with English ideas of propriety. The Duke's efforts at conciliatory manners, as evinced through much hand-shaking and baby-kissing, were not without significance. It was small matter for wonder, therefore, that the terrified authorities, before they got him out of their town, heartily regretted that they had not allowed him to have his own way after the accident to Mr. Huskisson, when he proposed to turn back without coming to it. Having once got him safely started back to Liverpool, therefore, they preferred to leave the other guests to take care of themselves, rather than have the Duke face the crowd again. As there were no sidings on that early road, and the connections between the tracks had, as a measure of safety, been temporarily removed, the ministerial train in moving towards Liverpool had necessarily shoved before it the engines belonging to the other trains. The unfortunate guests on those other trains, thus left to their fate, had for the rest of the day a very dreary time of it. To avoid accidents, the six trains abandoned at Manchester were united into one, to which were attached the three locomotives remaining. In this form they started. Presently the strain broke the couplings. Pieces of rope were then put in requisition, and again they got in motion. In due time the three other engines came along, but they could only be used by putting them on in front of the three already attached to the train. Two of them were used in that way, and the eleven cars thus drawn by five locomotives, and preceded at a short distance by one other, went on towards Liverpool. It was dark, and to meet the exigencies of the occasion the first germ of the present elaborate system of railroad night signals was improvised on the spot. From the foremost and pioneer locomotive obstacles were signalled to the train locomotives by the very primitive expedient of swinging the lighted end of a tar-rope. At Rainhill the weight of the train proved too much for the combined motive-power, and the thoroughly wearied passengers had to leave their carriages and walk up the incline. When they got to the summit and resuming their seats, were again in motion, fresh delay was occasioned by the leading locomotive running into a wheel-barrow, maliciously placed on the track to obstruct it. Not until ten o'clock did they enter the tunnel at Liverpool. Meanwhile all sorts of rumors of general disaster had for hours been circulating among the vast concourse of spectators who were assembled waiting for their friends, and whose relief expressed itself in hearty cheers as the train at last rolled safely into the station.

We have also Miss Kemble's story of this day, to which in her letter of August 25th she had looked forward with such eager interest. With her father and mother she had been staying at a country place in Lancashire, and in her account of the affair written in 1876 she says:

"The whole gay party assembled at Heaton, my mother and myself included, went to Liverpool for the opening of the railroad. The throng of strangers gathered there for the same purpose made it almost impossible to obtain a night's lodging for love or money; and glad and thankful were we, to put up with and be put up in a tiny garret by an old friend, Mr. Redley, of the Adelphi, which many would have given twice what we paid to obtain. The day opened gloriously, and never was an innumerable concourse of sight-seers in better humor than the surging, swaying crowd that lined the railroad with living faces. . . . After this disastrous event [the accident to Mr. Huskisson,] the day became overcast, and as we neared Manchester the sky grew cloudy and dark, and it began to rain. The vast concourse of people who had assembled to witness the triumphant arrival of the successful travelers was of the lowest order of mechanics and artisans, among whom great distress and a dangerous spirit of discontent with the government at that time prevailed. Groans and hisses greeted the carriage, full of influential personages, in which the Duke of Wellington sat. High above the grim and grimy crowd of scowling faces a loom had been erected, at which sat a tattered, starved-looking weaver, evidently set there as a *representative man*, to protest against this triumph of machinery, and the gain and glory which the wealthy Liverpool and Manchester men were likely to derive from it. The contrast between our departure from Liverpool and our arrival at Manchester was one of the most striking things I ever witnessed. * * *

MY DEAREST H—: MANCHESTER, *September* 20th, 1830.

* * * * * * *

" You probably have by this time heard and read accounts of the opening of the railroad, and the fearful accident which occurred at it, for the papers are full of nothing else. The accident you mention did occur, but though the unfortunate man who was killed bore Mr. Stephenson's name, he was not related to him. [Besides Mr. Huskisson, another man named Stephenson had about this time been killed on the railroad.] I will tell you something of the events on the fifteenth, as, though you may be acquainted with the circumstances of poor Mr. Huskisson's death, none but an eye-witness of the whole scene can form a conception of it. I told you that we had had places given to us, and it was the main purpose of our returning from Birmingham to Manchester to be present at what promised to be one of the most striking events in the scientific annals of our country. We started on Wednesday last, to the number of about eight hundred people, in carriages constructed as I before described to you. The most intense curiosity and excitement prevailed, and though the weather was uncertain, enormous masses of densely packed people lined the road, shouting and waving hats and handkerchiefs as we flew by them. What with the sight and sound of these cheering multitudes and the tremendous velocity with which we were borne past them, my spirits rose to the true champagne height, and I never enjoyed anything so much as the first hour of our progress. I had been unluckily separated from my mother in the first distribution of places, but by an exchange of seats which she was enabled to make she rejoined me, when I was at the height of my ecstacy, which was considerably damped by finding that she was frightened to death, and intent upon nothing but devising means of escaping from a situation which appeared to her to threaten with instant annihilation herself and all her travelling companions. While I was chewing the cud of this disappointment, which was rather bitter, as I expected her to be as delighted as myself with our excursion, a man flew by us, calling out through a speaking trumpet to stop the engine, for that somebody in the directors' car had sustained an injury. We were all stopped accordingly and presently a hundred voices were heard exclaiming that Mr. Huskisson was killed. The confusion that ensued is indescribable; the calling out from carriage to carriage to ascertain the truth, the contrary reports which were sent back to us, the hundred questions eagerly uttered at once, and the repeated and urgent demands for surgical assistance, created a sudden turmoil that was quite sickening. At last we distinctly ascertained that the unfortunate man's thigh was broken.

"From Lady W—, who was in the duke's carriage, and within three yards of the spot where the accident happened, I had the following details, the horror of witnessing which we were spared through our situation behind the great carriage. The engine had stopped to take in a supply of water, and several of the gentlemen in the directors' carriage had jumped out to look about them. Lord W—, Count Batthyany, Count Matuscenitz, and Mr. Huskisson among the rest were standing talking in the middle of the road, when an engine on the other line, which was parading up and down merely to show its speed, was seen coming down upon them like lightning. The most

active of those in peril sprang back into their seats ; Lord W—
saved his life only by rushing behind the duke's carriage, Count
Matuscenitz had but just leaped into it, with the engine all but
touching his heels as he did so; while poor Mr. Huskisson, less
active from the effects of age and ill health, bewildered too by
the frantic cries of 'Stop the engine! Clear the track!' that
resounded on all sides, completely lost his head, looked help-
lessly to the right and left, and was instantaneously prostrated
by the fatal machine, which dashed down like a thunderbolt
upon him, and passed over his leg, smashing and mangling it in
the most horrible way. (Lady W— said she distinctly heard
the crushing of the bone.) So terrible was the effect of the ap-
palling accident that except that ghastly "crushing" and poor
Mrs. Huskisson's piercing shriek, not a sound was heard or a
word uttered among the immediate spectators of the catastrophe.
Lord W— was the first to raise the poor sufferer, and calling to
his aid his surgical skill, which is considerable, he tied up the
severed artery, and for a time at least, prevented death by a loss
of blood. Mr. Huskisson was then placed in a carriage with his
wife and Lord W--, and the engine having been detached from
the directors' carriage, conveyed them to Manchester. So great
was the shock produced upon the whole party by this event that
the Duke of Wellington declared his intention not to proceed,
but to return immediately to Liverpool. However, upon its
being represented to him that the whole population of Manchester had turned out to witness the procession, and that a disappointment might give rise to riots and disturbances, he con-
sented to go on, and gloomily enough the rest of the journey
was accomplished. We had intended returning to Liverpool by
the railroad, but Lady W—, who seized upon me in the midst
of the crowd, persuaded us to accompany her home, which we
gladly did. Lord W—, did not return till past ten o'clock, at
which hour he brought the intelligence of Mr. Huskisson's
death. I need not tell you of the sort of whispering awe which
this event threw over our whole circle; and yet great as was the

horror excited by it, I could not help feeling how evanescent the effect of it was, after all. The shuddering terror of seeing our fellow-creature thus struck down by our side, and the breathless thankfulness for our own preservation, rendered the first evening of our party at Heaton almost solemn; but the next day the occurrence became a subject of earnest, it is true, but free discussion; and after that was alluded to with almost as little apparent feeling as if it had not passed under our eyes, and within the space of a few hours."

In spite of accidents and *contre-temps*, however, the road was opened to traffic, and at once proceeded to outdo in its results the most eager anticipations of its friends. No account of its first beginnings would, however, be complete for our time, which did not also give an idea of the impressions produced on one travelling over it before yet the novelty of the thing had quite worn away. It was a long time, comparatively, after September, 1830, before the men who had made a trip over the railroad ceased to be objects of deep curiosity. Here is the account of his experience by one of these far-travelled men, with all its freshness still lingering about it:

"Although the whole passage between Liverpool and Manchester is a series of enchantments, surpassing any in the Arabian Nights, because they are realities, not fictions, yet there are certain epochs in the transit which are peculiarly exciting. These are the startings, the ascents, the descents, the tunnels, the Chat Moss, the meetings. At the instant of starting, or rather before, the automaton belches forth an explosion of steam, and seems for a second or two quiescent. But quickly the explosions are reiterated, with shorter and shorter intervals, till they become too rapid to be counted, though still distinct.

2*

These belchings or explosions more nearly resemble the pantings of a lion or tiger, than any sound that has ever vibrated on my ear. During the ascent they became slower and slower, till the automaton actually labors like an animal out of breath, from the tremendous efforts to gain the highest point of elevation. The progression is proportionate; and before the said point is gained, the train is not moving faster than a horse can pace. With the slow motion of the mighty and animated machine, the breathing becomes more laborious, the growl more distinct, till at length the animal appears exhausted, and groans like the tiger, when overpowered in combat by the buffalo.

"The moment that the height is reached and the descent commences, the pantings rapidly increase; the engine with its train starts off with augmenting velocity; and in a few seconds it is flying down the declivity like lightning, and with a uniform growl or roar, like a continuous discharge of distant artillery.

"At this period, the whole train is going at the rate of thirty-five or forty miles an hour! I was on the outside, and in front of the first carriage, just over the engine. The scene was magnificent, I had almost said terrific. Although it was a dead calm the wind appeared to be blowing a hurricane, such was the velocity with which we darted through the air. Yet all was steady; and there was something in the precision of the machinery that inspired a degree of confidence over fear—of safety over danger. A man may travel from the Pole to the Equator, from the Straits of Malacca to the Isthmus of Darien, and he will see nothing so astonishing as this. The pangs of Etna and Vesuvius excite feelings of horror as well as of terror; the convulsion of the elements during a thunderstorm carries with it nothing but pride, much less of pleasure, to counteract the awe inspired by the fearful workings of perturbed nature; but the scene which is here presented, and which I cannot adequately describe, engenders a proud consciousness of superiority in human ingenuity, more intense and convincing than any effort or product of the poet, the painter, the philosopher, or the divine. The projections or transits of the train through the tunnels or arches, are very electrifying. The deafening peal of thunder, the sudden immersion in gloom, and the clash of reverberated sounds in confined space, combine to produce a momentary shudder or idea of destruction—a thrill of annihilation, which is instantly dispelled on emerging into the cheerful light.

"The meetings or crossings of the steam trains flying in opposite directions are scarcely less agitating to the nerves than their transits through the tunnels. The velocity of their course, the propinquity or apparent identity of the iron orbits along which these meteors move, call forth the involuntary but fearful thought of a possible collision, with all its horrible consequences. The period of suspense, however, though exquisitely painful, is but momentary; and in a few seconds the object of terror is far out of sight behind.

"Nor is the rapid passage across Chat Moss unworthy of notice. The ingenuity with which two narrow rods of iron are made to bear whole trains of wagons, laden with many hundred tons of commerce, and bounding across a wide, semi-fluid morass, previously impassable by man or beast, is beyond all praise and deserving of eternal record. Only conceive a slender bridge of two minute iron rails, several miles in length, level as Waterloo, elastic as whalebone, yet firm as adamant! Along this splendid triumph of human genius—this veritable *via triumphalis*—the train of carriages bounds with the velocity of the stricken deer; the vibrations of the resilient moss causing the ponderous engine and its enormous suite to glide along the surface of an extensive quagmire as safely as a practiced skater skims the icy mirror of a frozen lake.

"The first class or train is the most fashionable, but the second or third are the most amusing. I travelled one day from Liverpool to Manchester in the lumber train. Many of the carriages were occupied by the swinish multitude, and others by a multitude of swine. These last were naturally vociferous if not eloquent. It is evident that the other passengers would have been considerably annoyed by the orators of this last group, had there not been stationed in each carriage an officer somewhat analogous to the Usher of the Black Rod, but whose designation on the railroad I found to be 'Comptroller of the Gammon.' No sooner did one of the long-faced gentlemen raise his note too high, or wag his jaw too long, than the 'Comptroller of the Gammon' gave him a whack over the snout with the butt end of his shillelagh; a snubber which never failed to stop his oratory for the remainder of the journey."

To one familiar with the history of English railroad legislation the last paragraph is peculiarly significant. For years after the railroad system was inaugurated, and until legislation was invoked to compel something better, the companies persisted in carrying passengers of the third class in uncovered carriages, exposed to all weather, and with no more decencies or comforts than were accorded to swine.

Naturally, the beginning of the railroad system in America, was neither so interesting nor so picturesque as it had been in the case of Great Britain. At most it was but an imitation; and that too, on a small scale. Yet, about all its details there was something which cannot but be peculiarly suggestive to the American of the present day. As you review the record, it seems to relate to another country and almost to a different world. With the Manchester & Liverpool road this was not so. There the thing, for a beginning, was on a large scale. The cost of the structure, the number of the locomotives, the fame of the guests, the mass and excitement of the spectators were all equal to the occasion. This was not so in America. Everything was diminutive and poor in 1831. The provincialism of the time and place is almost oppressive. In turning over the old records the eye constantly rests on the names, familiar to us, of men now living; but it seems scarcely possible that any human life can have spanned the well nigh incredible gap which separates the America of 1878 from that of 1830. Certainly, neither anywhere else nor at any other time has the world in a space of less than fifty years witnessed such extraordinary development.

Whatever credit is due to the construction of the first railroad ever built in America is usually claimed for the State of Massachusetts. Every one who has ever looked into a school history of the United States knows something ôf the Quincy railway of 1826. Properly speaking, however, this was never—or at least, never until the year 1871,—a railroad at all. It was nothing but a specimen of what had been almost from time immemorial in common use in England, under the name of "tram-ways." Indeed it is a curious illustration of the combined poverty and backwardness of America at that time, that so common and familiar an appliance should only then have been introduced, and should have excited so much interest and astonishment. This road, known as the Granite railway, was built by those interested in erecting the

Bunker Hill Monument, for the purpose of getting the stone down from the Quincy quarries to a wharf on Neponset River, from which it was shipped to its destination. The whole distance was three miles, and the cost of the road was about $34,000. At the quarry end there was a steep inclined plane, up and down which the cars were moved by means of a stationary engine. From the foot of that incline the road sloped gently off to its river terminus. There was nothing in its construction which partook of the character of a modern railroad. The tracks were five feet apart, and laid on stone sleepers eight feet apart. On this stone substructure wooden rails were laid, and upon these another rail of strap iron. Down this road two horses could draw a load of forty tons, and thus the expense of moving stone from the quarries to the river was reduced to about a sixth part of what it was while the highway alone was in use.

Such was the famous Quincy railway, the construction of which is still referred to as marking an era of the first importance in American history. Such, also, it remained down to the year 1871,—a mere tramway, operated exclusively by means of horses. In that year the franchise was at last purchased by the Old Colony Railroad Company, the ancient structure was completely demolished, and a modern railroad built on the right of way. Through the incorporation into it of the old Granite railway, therefore, the line which connects the chief town of what was once Plymouth Colony with the chief town of what was once the colony of Massachusetts Bay has become the oldest railroad line in America. In this there is, so to speak, a manifest historical propriety.

Apart, however, from the construction of the Granite railway, Massachusetts was neither particularly early nor particularly energetic in its railroad development. At a later day many of her sister States were in advance of her, and especially was this true of South Carolina. There is, indeed, some reason for believing that the South Carolina Railroad was the first ever constructed in any country with a definite plan of operating it exclusively by locomotive steam power. But in America there was not,—indeed from the very circumstances of the case there could not have been,—any such dramatic occasions and surprises as those witnessed at Liverpool in 1829 and 1830. Nevertheless the people of Charleston were pressing close on the heels of those of Liverpool, for on the 15th of January 1831,—exactly four months after the formal opening of the Manchester & Liverpool road,—the first anniversary of the South Carolina Railroad was celebrated with due honor. A queer looking machine, the outline of which was sufficient in itself to prove that the inventor owed nothing to Stephenson, had been constructed at the West Point Foundry Works in New York during the summer of 1830—a first attempt to supply that locomotive power which the Board had, with a sublime confidence in possibilities, unanimously voted on the 14th of the preceding January should alone be used on the road. The name of *Best Friend* was given to this very simple product of native genius. The idea of the multitubular boiler had not yet suggested itself in America. The *Best Friend*, therefore, was supplied with a common vertical boiler " in form of an old fashioned porter-bottle, the furnace at the bottom surrounded with water, and all filled inside of what we call teats, running out from the sides and tops." By means of these projections, or " teats," a portion at least of the necessary heating surface was provided. The cylinder was at the front of the platform, the rear end of which was occupied by the boiler, and it was fed by means of a connecting pipe. Thanks to the indefatigable researches of an enthusiast on railroad construction, we have an account of the performances of this, and all the other pioneers among American locomotives; and the pictures with which Mr. W. H. Brown has enriched his book* would alone render it both curious and valuable. Prior to the stockholders' anniversary of January 15th, 1831, it seems that the *Best Friend* had made several trial trips " running at the rate of sixteen to twenty-one miles an hour, with forty or fifty passengers in some four or five cars, and without the cars, thirty to thirty-five miles per hour." The stockholders' day was, however, a special occasion, and the papers of the following Monday, for it happened on a Saturday, gave the following account of it:

* The History of the First Locomotive in America.

"Notice having been previously given, inviting the stockholders, about one hundred and fifty assembled in the course of the morning at the company's building in Line Street, together with a number of invited guests. The weather the day and night previous had been stormy, and the morning was cold and cloudy. Anticipating a postponement of the ceremonies, the locomotive engine had been taken to pieces for cleaning, but upon the assembling of the company she was put in order, the cylinders new packed, and, at the word, the apparatus was ready for movement. The first trip was performed with two pleasure-cars attached, and a small carriage, fitted for the occasion, upon which was a detachment of United States troops and a field piece which had been politely granted by Major Belton for the occasion. . . The number of passengers brought down, which was performed in two trips, was estimated at upward of two hundred. A band of music enlivened the scene, and great hilarity and good humor prevailed throughout the day."

The "great hilarity and good humor" of this occasion no one can doubt who studies the supposed contemporaneous picture of it contained in Mr. Brown's book. The pleasure must, however, have been largely due to novelty, inasmuch as a railroad journey on a "cold and cloudy" January day, performed in "two pleasure cars," between which and an "old fashioned porter-bottle" of a locomotive, puffing out smoke and cinders, there was nothing but a "small carriage" fitted up to carry "a field piece," while a band of music enlivened the whole—taking all these ingredients together, it would not at this time seem easy to compound from them a day of high physical enjoyment. But the fathers were a race of simpler tastes.

It was not long, however, before the *Best Friend*

came to serious grief. Naturally, and even necessarily, inasmuch as it was a South Carolina institution, it was provided with a negro fireman. It so happened that this functionary while in the discharge of his duties was much annoyed by the escape of steam from the safety-valve, and, not having made himself complete master of the principles underlying the use of steam as a source of power, he took advantage of a temporary absence of the engineer in charge to effect a radical remedy of this cause of annoyance. He not only fastened down the valve lever, but further made the thing perfectly sure by sitting on it. The consequences were hardly less disastrous to the *Best Friend* than to the chattel fireman. Neither were of much further practical use. Before this mishap chanced, however, in June, 1831, a second locomotive, called the *West Point*, had arrived in Charleston; and this at last was constructed on the principle of Stephenson's *Rocket*. In its general aspect, indeed, it greatly resembled that already famous prototype. There is a very characteristic and suggestive cut representing a trial trip made with this locomotive on March 5th, 1831. The nerves of the Charleston people had been a good deal disturbed and their confidence in steam as a safe motor shaken by the disaster which had befallen the *Best Friend*. Mindful of this fact, and very properly solicitous for the safety of their guests, the directors now had recourse to a very simple and ingenious expedient. They put what they called a " barrier car " between the locomotive and passenger coaches of the train. This barrier car consisted of a platform on wheels upon which were piled six bales of cotton. A fortification was thus provided between the passengers and any future negro sitting on the safety valve. We are also assured that "the safety valve being out of the reach of any person but the engineer, will contribute to the prevention of accidents in future, such as befel the *Best Friend*. Judging by the cut which represents the train, this occasion must have been even more marked for its "hilarity" than the earlier one which has already been described. Besides the locomotive and the barrier car there are four passenger coaches. In the first of these was a negro band, in general appearance very closely resembling the minstrels of a later day, the members of which are energetically performing on musical instruments of various familiar descriptions. Then follow three cars full of the saddest possible looking white passengers, who were present as we are informed to the number of one hundred and seventeen. The excursion was, however, highly successful, and two and a quarter miles of road were passed over in the short space of eight minutes,—about the speed at which a good horse would trot for the same distance.

This was in March, 1831. About six months before, however, there had actually been a trial of speed between a horse and one of the pioneer locomotives, which had not resulted in favor of the locomotive. It took place on the present Baltimore & Ohio road upon the 28th of August, 1830. The engine in this case was contrived by no other than Mr. Peter Cooper. And it affords a striking illustration of how recent those events which now seem so remote really were, that here is a man still living, and among the most familiar to the eyes and mouths of the present generation, who was a contemporary of Stephenson, and himself invented a locomotive during the Rainhill year, being then nearly forty years of age. The Cooper engine, however, was scarcely more than a working model. Its active-minded inventor hardly seems to have aimed at anything more than a demonstration of possibilities. The whole thing weighed only a ton, and was of one-horse power; in fact it was not larger than those hand-cars now in common use with railroad section-men. The boiler, about the size of a modern kitchen boiler, stood upright and was filled above the furnace, which occupied the lower section, with vertical tubes. The cylinder was but three and a half inches in diameter, and the wheels were moved by gearing. In order to secure the requisite pressure of steam in so small a boiler, a sort of bellows was provided which was kept in action by means of a drum attached to one of the car-wheels over which passed a cord which worked a pulley, which in turn worked the bellows. Thus of Stephenson's two great devices, without either of which his success at Rainhill would have been impossible,—the waste-steam blast and the multitubular boiler,—Peter Cooper had only got hold of the last. He owed his defeat in the race between his engine and a horse to the fact that he had not got hold of the first. It happened in this wise. Several experimental trips had been made with the little engine on the Baltimore & Ohio road, the first sections of which had recently been completed and were then operated by means of horses. The success of these trips was such, that at last, just seventeen days before the formal opening of the Manchester & Liverpool road on the other side of the Atlantic, a small open car was attached to the engine,—the name of which, by the way, was *Tom Thumb*—and upon this a party of directors and their friends were carried from Baltimore to Ellicott's Mills and back; a distance of some twenty-six miles. The trip out was made in an hour and was very successful. The return was less so, and for the following reason:

" The great stage proprietors of the day were Stockton and Stokes; and on that occasion a gallant gray, of great beauty and power, was driven by them from town, attached to another car on the second track—for the company had begun by making two tracks to the Mills—and met the engine at the Relay House, on its way back. From this point it was determined to have a race home, and the start being even, away went horse and engine, the snort of the one and the puff of the other keeping tune and time.

" At first the gray had the best of it, for his *steam* would be applied to the greatest advantage on the instant, while the engine had to wait until the rotation of the wheels set the blower to work. The horse was perhaps a quarter of a mile ahead, when the safety valve of the engine lifted, and the thin blue vapor issuing from it showed an excess of steam. The blower whistled, the steam blew off in vapory clouds, the pace increased, the passengers shouted, the engine gained on the horse, soon it lapped him—the silk was plied—the race was neck and neck, nose and nose,—then the engine passed the horse, and a great hurrah hailed the victory. But it was not repeated, for just at this time, when the gray's master was about giving up, the band which draws the pulley which moved the blower slipped from the drum, the safety valve ceased to scream, and the engine, for want of breath, began to wheeze and pant. In vain Mr. Cooper, who was his own engineer and fireman, lacerated his hands in attempting to replace the band upon the wheel; the horse gained on the machine and passed it, and although the band was presently replaced, and steam again did its best, the horse was too far ahead to be overtaken, and came in the winner of the race."

Poor and crude as the country was, however, America showed itself far more ready to take in the far reaching consequences of the initiative which Great Britain gave in 1830 than any other country in the world. Belgium, under the enlightened rule of King Leopold, did not move in the new departure until 1834, and France was slower yet. The fact is, however, that those countries did not feel the need of the railroad at all in the same degree as either England or America. They already had excellent systems of roads which sufficed for all their present needs. In America, on the contrary, the roads were few and badly built; while in England, though they were good enough, the volume of traffic had outgrown their capacity. America suffered from too few roads; England from too much traffic. Both were restlessly casting about for some form of relief. Accordingly, all through the time during which Stephenson was fighting the battle of the locomotive, America, as if in anticipation of his victory, was building railroads. It might almost be said that there was a railroad mania. Massachusetts led off in 1826; Pennsylvania followed in 1827, and in 1828 Maryland and South Carolina. Of the great trunk lines of the country, a portion of the New York Central was chartered in 1825; the construction of the Baltimore & Ohio was begun on July 4th, 1828. The country, therefore, was not only ripe to accept the results of the Rainhill contest, but it was anticipating them with eager hope. Had George Stephenson known what was going on in America he would not when writing to his son in 1829 have limited his anticipation of orders for locomotives to "at least thirty."

Accordingly, after 1830 trial trips with new locomotive engines followed hard upon each other. To-day it was the sensation in Charleston; to-morrow in Baltimore; the next day at Albany. Reference has already been made to a cut representing the excursion train of March 5th, 1831, on the South Carolina Railroad. There is, however, a much more familiar picture of a similar trip made on the 9th of August of the same year from Albany to Schenectady, over the Mohawk Valley road. This sketch, moreover, was made at the time and on the spot by Mr. W. H. Brown, whose book has already been referred to. There are few things of the sort more familiar to the general eye, and, either in shop windows or in the offices of railroad companies, almost every one has curiously studied the train, with its snorting little engine and barrels of pine-knots for fuel,—the highly respectable looking engineer, standing up and meditatively observing his machine, with all the dignity inseparable from a dress coat so neatly buttoned up,—the two following coaches, in the inside of the first of which may be identified Mr. Thurlow Weed, already a man of thirty years of age and one of the political powers of the land, while he upon the outside seat, with his hat on the back of his head, is, as we are informed, Mr. Billy Winn, the penny post man. The history of this now famous excursion has been preserved almost as minutely as that of the more widely known affair which had taken place at Liverpool just a year before. The train was made up of a locomotive, the *De Witt Clinton*, its tender and five or six passenger coaches, which were, indeed, nothing but the bodies of stage coaches placed upon trucks. The first two of these coaches were set aside for distinguished visitors; the others were surmounted with seats of plank to accommodate as many as possible of the great throng of persons who were anxious to participate in the trip. Inside and out the coaches were crowded; every seat was full. At Liverpool the start of the train was signalled by the discharge of a cannon; they were more modest at Albany, where the conductor, having duly collected his tickets by stepping from platform to platform outside the cars, mounted on the tender and, sitting upon the little seat which is to be seen in the sketch of the train, gave the signal to start with a tin horn. What followed has been described by one who took part in the affair:

"The trucks were coupled together with chains or chain-links, leaving from two to three feet slack, and when the locomotive started it took up the slack by jerks, with sufficient force to jerk the passengers, who sat on seats across the tops of the coaches, out from under their hats, and in stopping they came together with such force as to send them flying from their seats.

"They used dry pitch-pine for fuel, and there being no smoke or spark-catcher to the chimney or smoke-stack, a volume of black smoke, strongly impregnated with sparks, coal and cinders, came pouring back the whole length of the train. Each of the outside passengers who had an umbrella raised it as a protection against the smoke and fire. They were found to be but a momentary protection, for I think in the first mile the last one went overboard, all having their covers burnt off from the frames, when a general mêlée took place among the deck-passengers, each whipping his neighbor to put out the fire. They presented a very motley appearance on arriving at the first station." Here "a short stop was made, and a successful experiment tried to remedy the unpleasant jerks. A plan was soon hit upon and put into execution. The three links in the couplings of the cars were stretched to their utmost tension, a rail, from a fence in the neighborhood, was placed between each pair of cars and made fast by means of the packing yarn from the cylinders. This arrangement improved the order of things, and it was found to answer the purpose when the signal was again given and the engine started."

In spite of these trifling annoyances the engine, which was a little thing weighing but three and a half tons, accomplished the distance from Albany to Schenectady in less than an hour, and, during a part of the way, ran at as high a speed as a mile in two minutes. At Schenectady the members of the party refreshed themselves, and then, resuming their seats, reached Albany in due time and without delay or accident of any kind. In spite of dilapidated garments and lost umbrellas, the passengers were on the whole well pleased with their trip, and in this respect at any rate far more fortunate than those who a year before had helped inaugurate the Manchester & Liverpool road.

The *DeWitt Clinton*, as well as all the other engines used on the occasions which have been described, were of American make. But the fame of the Stephenson works at Newcastle-upon-Tyne had crossed the ocean, and to possess a specimen of their products was the ambition of every enterprising railroad company. As early as September 1829 one of their earlier engines,—the *Stourbridge Lion* by name—had been landed in New York and set up as an object of curiosity in an iron-yard on the East River. It was one of the old models, however, of ante-Rocket construction, and was always regarded as a failure. Orders for locomotives of the new model were sent over as soon as its success was demonstrated, and at about the time of the Schenectady excursion, which has just been described, one of these was landed in New York. Its whole cost, including freight, duties and insurance, was $4,869.59. This "powerful Stephenson locomotive," as it was called, weighed about seven tons; but, light as this now seems, it was far too heavy for the structure upon which it was to run. Early in September, however, it was placed upon the tracks, and on the 16th and 17th made its trial trips, using coal for fuel. And now another, and this time more formal and brilliant excursion, was planned in honor of the event, and was fixed for the 24th. It was to be to America what the Manchester & Liverpool opening had been to England, and the presence of the Duke of Wellington there, was to be offset here by that of Gen. Scott. But to be appreciated this excursion must be described in contemporaneous language:

"The company consisted of the Governor, Lieutenant Governor, members of the Senate, now in session as a Court of Errors, our Senators in Congress, the Chancellor and Judges of the Supreme and District Courts, State officers, the president of the Board of Assistants and members of the Common Council of the city of New York, the Mayor, Recorder and Corporation Counsel of the city, and several citizens of New York, Albany and Schenectady.

"Owing to a defect in one of the supply-pipes of the English locomotive, that powerful engine was not brought into service, and the party, having been delayed in consequence, did not leave the head of Lydius Street until nearly twelve o'clock. They then started with a train of ten cars, three drawn by the American locomotive, *De Witt Clinton,* and seven by a single horse each. The appearance of this fine cavalcade, if it may be so called, was highly imposing. The trip was performed by the locomotive in forty-six minutes, and by the cars drawn by horses in about an hour and a quarter. From the head of the plane, about a quarter of a mile from Schenectady, the company were conveyed in carriages to Davis's Hotel, where they were joined by several citizens of Schenectady, and partook of a dinner that reflected credit upon the proprietor of that well known establishment. Among the toasts offered was one which has been verified to the letter, viz. : ' The Buffalo Railroad—may we soon breakfast in Utica, dine in Rochester, and sup with our friends on Lake Erie!" After dinner the company repaired to the head of the plane, and resumed their seats for the return to Albany. It was an imposing spectacle."

All this took place during the summer of 1831, and it was only during that very summer, when the locomotive was already an established fact, a working agent in at least two of her sister States, that Massachusetts aroused herself to a consciousness that something unusual had taken place. Then at last that corporation was chartered which subsequently, five years later, opened for public use the first steam railroad regularly planned and constructed as such within the limits of the State. Such an apparent apathy is not very explicable. Not only the Quincy railway, but the Middlesex canal as well, had been the first things of the kind brought to a successful completion in America. Just at this time, also, the manifest success of the Erie canal had given a new and portentous significance to the Berkshire hills, causing them to throw a dark shadow over the future of Massachusetts. They seemed stationed on the western border of the State, an insuperable barrier against which the eastward tide of commerce struck and then with a deflected course flowed quietly in the direction of New York. Either in some way that barrier must be overcome or the material progress of the State would in the future be seriously threatened. So much was obvious. Before the year 1826 this difficult problem had already occupied the attention both of the public and of the legislature. A commission had been appointed to survey a canal route from tide water at Boston to the Connecticut, and thence to some point in the State of New York, near where the Erie canal emptied into the Hudson. The report of this commission was submitted by Gov. Lincoln to the legislature in January 1826, and to-day, after the lapse of more than fifty years, the document has a peculiar interest and significance.

The survey was made by Col. Loammi Baldwin, a civil engineer who has left his mark cut deep on the Massachusetts system of internal improvements. There was no good route, and so he fixed on what has since become well known as the Hoosac Tunnel line as being the least bad. Accompanying the report of 1826 is a map made by Col. Baldwin upon which is laid down a canal-tunnel exactly where the railroad tunnel now is. This canal-tunnel project was adopted by Col. Baldwin as a dreadful alternative to a system of locks crossing the mountains at the same point. Not that he considered the lock scheme impracticable; on the contrary, he demonstrated in his report its perfect feasibility on paper He objected to it solely on the score of expense. He accordingly had recourse to the cheaper expedient of a tunnel, and proceeded to estimate its cost. That long forgotten estimate is now one of the curiosities of engineering literature. It was made, be it remembered, in the early days before tunnelling had become a science, and when the whole work would necessarily have been done by hand-drilling and without the aid of any explosive more powerful than gunpowder. In making his estimate Col. Baldwin, as an engineer of character having a reputation at stake, was extremely cautious. He said, "in a tunnel, four miles in length, of the size named, there will be 211,200 cubic yards of stone to excavate, which at $4.25 per cubic yard, amounts to $920,832." But this he took pains to state was "beyond a doubt; the highest price" having been assumed. Even at the time, this conclusion, to which subsequent bitter experience has lent a grim humor, did not pass unchallenged. A writer in the Boston *Courier*, for instance, calculated that, on the data given in the report, it would take fifty-two years and nineteen days to finish the tunnel. The present Hoosac Tunnel was in fact finished a little over fifty years from the time when this report was laid before the Legislature; but, instead of having proved "not more difficult than the cut through the 'Mountain Ridge' on the Erie canal," the expense of which per cubic yard had been $1.75, each new difficulty which developed itself was overcome only to make way for another, until the ultimate expense was about $20 per cubic yard, and the total cost some ten times the original estimate.

Naturally enough, however, nothing was done in

consequence of Col. Baldwin's report towards extending the Erie canal to a connection with tide-water at Boston. That such an idea should ever have been gravely entertained seems now almost beyond belief. Yet there, on file among the public documents of that day, is the record showing that sane men actually dreamed that water could be made to flow over the Berkshire hills so as to compete with the untaxed current of the Hudson! Four years more passed by without contributing anything to the solution of the problem. In October 1829, however, the crucial test at Rainhill gave a new direction to men's thoughts in other places than in England. Nathan Hale at that time edited the Boston *Daily Advertiser*, and he had also been one of the commissioners under whom Col. Baldwin had made his survey. An editor of a school which has long since passed away, he not only occupied a prominent position in the business circles of the day, but by force of individual character he exercised through his paper a wide and useful influence. The *Advertiser* was Nathan Hale; and, as regarded this question, Nathan Hale moved in the front rank of progress. And now in 1829 the *Advertiser* reproduced day after day every detail of the Rainhill experiments and spread them before the people of Massachusetts with all possible emphasis. The result was immediate. The sessions of the Massachusetts legislature were then held in the spring and early summer, and occupied about as many weeks as they now do months. The following June, in response as it were to the Rainhill challenge and before the Manchester & Liverpool road was yet open to traffic, a number of charters were granted to corporations with all necessary powers to construct railroads on specified routes. In those days, however, the science of railroad financiering had not been developed to the degree of excellence which it has since attained. The first condition to an enterprise was the actual raising of money to carry it out. While capitalists were timid, legislatures were cautious. A tedious contest, therefore, ensued between the two. On the one hand, those who proposed to build the roads insisted on a guarantee of exclusive railroad rights between the points designated in their charters; on the other hand, the legislature refused to concede any such monopoly. At last, however, in the case of a single road, that between Boston and Lowell, the exclusive concession asked for was granted for a term of years. Accordingly in this case the charter was accepted, the company speedily organized and books of subscription opened.

In November 1830 accounts came across the ocean of the successful opening of the Manchester & Liverpool road, and again Mr. Hale reproduced them in detail. The schoolmaster was abroad. He was busily at work between Liverpool and Manchester, and the *Advertiser* kept his instructions before the eyes of the people at home. Consequently, in June 1831, two more roads were incorporated by the next legislature, one of which, the Boston & Providence, still preserves its individuality and its original name. The boards of direction of these several roads were made up of the most respectable and best known citizens of Boston. Patrick T. Jackson was president of the Lowell company; T. B. Wales of the Providence, and Nathan Hale of the Worcester. There is something very interesting and attractive in those days of small things. To men of the present time, accustomed to corporations which operate thousands of miles of road, which yearly carry millions of tons of freight and tens of millions of passengers, while they wield hundreds of millions of capital,—to men accustomed to the presence of these leviathans, the little original roads, the longest of which was but fifty miles, seem little more than toys. They were, however, the beginning of great things. We to-day are familiar with the names of enterprises which stretch out into what was then the undiscovered West, and the fabulous East. We can, whenever we please, read the last quotation of stocks representing a property lying on the shores of the Euphrates or among the steppes and gorges of the Rocky or Ural Mountains. We have tunnelled the Alps and bridged the Mississippi. These great accomplished facts, however, only make the fresh, new impressions with which our fathers viewed the gradual completion of the little original lines more quaint and more interesting. The gossip, as it were, of those days is by no means the least attractive thing about them.

The Lowell was the first organized of the Massachusetts roads, as well as the first upon which the work of construction was actually begun, though the Boston & Providence was the first completed. But it was upon the Worcester road, and towards the latter part of March, 1834, that the first locomotive ever used in Massachusetts was set in motion. On the 24th of the month Mr. Hale advised the readers of the *Advertiser* that "the rails are laid, from Boston to Newton, a distance of nine or ten miles, to which place it is proposed to run the passenger cars as soon as two locomotives shall be in readiness, so as to ensure regularity. One locomotive, called the *Meteor*, has been partially tried and will probably be in readiness in a few days; the second, called the *Rocket*, is waiting the arrival of the builder for subjecting it to a trial, and the third it is hoped will be ready by the first of May." The last named locomotive, the *Rocket*, was built by the Stephensons at Newcastle-upon-Tyne, and "the builder" whose arrival was looked for must have been an English engineer sent out to superintend the work of putting it in operation. No allusion is made in the papers to the first trial of these locomotives, but we have the impressions which one who claims to have been an eye-witness of it long afterwards gave:—

"The Boston & Worcester Railroad Company imported from Newcastle-upon-Tyne one of George Stephenson's locomotives, small in stature but symmetrical in every particular, and finished with the exactness of a chronometer. Placed upon the track, its driver, who came with it from England, stepped upon the platform with almost the airs of a juggler or a professor of chemistry, placed his hand upon the lever, and with a slight move of it the engine started at a speed worthy of the companion of the *Rocket* amid the shouts and cheers of the multitude. It gave me such a shock, that my hair seemed to start from the roots, rather than to stand on end."

On the 4th of April, a Friday by the way, a locomotive was first employed on a gravel train, upon which occasion, as the *Advertiser* the next day assured its readers, "the engine worked with ease, was perfectly manageable, and showed power enough to work at any desirable speed." Three days later, on Monday, 7th, we are informed that a locomotive ran on the railroad for the first time, " as far as Davis' tavern in Newton, a distance of eight or nine miles, accompanied by a part of the directors and fifty or sixty other persons, for the purpose of making trial of the engine and examination of the road. The party stopped several times for various purposes on the way out. They returned in thirty-nine minutes, including a stop of about six minutes for the purpose of attaching five cars loaded with earth. The engine travelled with ease at the rate of twenty miles an hour." The next day a larger party went over the ground, the directors inviting about one hundred and thirty gentlemen on the excursion. It would not appear to have been a very successful affair, for, "after proceeding a short distance, their progress was interrupted by the breaking of a connecting-rod between two of the cars. This accident caused a considerable delay, and unfortunately the same accident occurred three or four times during the excursion." So, after a short stop at Newton, the party came back, quite cross apparently, and did not get home until half past six in the evening. On the 15th of the month a yet larger party, consisting of about one hundred and twenty ladies and gentlemen in six cars, went out to Newton and back, making the return trip in less than half an hour. The cars began to run regularly next day, making two trips each way to Newton and back, leaving Boston at 10 A. M. and at 3.30 P. M. The regular passenger railroad service in Massachusetts dates, therefore, from the 16th of May 1834. Already, four days before, there had appeared in the advertising columns of Mr. Hale's paper a new form of notice. At the head was a rude cut of a locomotive and part of a train of cars,—the cars being of the old stage coach pattern, mounted high on wheels with spokes in them, and divided into compartments which were entered through doors at the sides. The brakemen, sitting on a sort of coach boxes, regulated the speed by the pressure of their feet on levers just as is still done with wagons on hilly roads. The notice was headed " Boston & Worcester Railroad " and read as follows :

"The passenger cars will continue to run daily from the depot near Washington St., to Newton, 6 o'clock and 10 o'clock A. M. and at $3\frac{1}{2}$ o'clock P. M., and

" Returning, leave Newton at 7, and a quarter past 11 A. M., and a quarter before 5 P. M.

"Tickets for the passage either way may be had at the Ticket Office, No. 617 Washington St., price thirty-seven and a half cents each; and for the return passage, of the Master of the cars, Newton.

"By order of the President and Directors.
"F. A. WILLIAMS, Clerk."

Curiously enough the issue of the very next day contains this editorial notice:

"*History of the United States.* We understand that George Bancroft Esq. of Northampton, has been for a long time engaged in the preparation of a history of the United States, the first volume of which is nearly ready for publication. Mr. Bancroft has been long known to the public as a scholar of distinguished talent, and diversified attainments; and there is every reason to believe that his work will be equally honorable to himself and to the literary reputation of the country. He has qualified himself for the task by a very diligent investigation of authorities, and a resort to the most authentic sources, in order to render his history no less valuable as a repertory of well ascertained facts, than it will be spirited and interesting as a narrative."

It would certainly be no easy matter to hit upon any incident which could more forcibly illustrate the briefness of the time within which the whole railroad development has been compressed. Mr. Bancroft is still at work on his history. The consecutive labor of one literary life time covers, therefore, the entire period. The later volumes of Bancroft's history are still unpublished; yet when the earliest railroad train was run in New England the first volume was just issuing from the press.

Partially opened to travel in May, the Boston &

Worcester road was by the end of June finished as far as Needham, and on the 7th of July it was formally opened to that point; when "the stockholders and a number of other gentlemen, to the number of about two hundred in all, by invitation of the directors, made an excursion to Needham, in eight passenger cars, drawn by the new locomotive *Yankee*... The excursion was pleasant, and the party appeared to enjoy the ride, and the beautiful scenery which is presented to view on different parts of the route." The further extension to Hopkinton was completed by September, and so on the 20th of that month another excursion of some two hundred in number went out from Boston in seven of the company's largest passenger cars drawn by the locomotive *Yankee*, and duly celebrated the occasion. "They started off" as the *Advertiser* of the following day stated, "at a rapid and steady pace. The weather was unusually fine, and the sweetness of the atmosphere, the rapidity of the motion, and the beauty and novelty of the scenery which was successively presented to view, appeared to produce in all the party an agreeable exhilaration of spirits." At Framingham the excursionists were met by John Davis, then Governor of the Commonwealth, by ex-Governor Lincoln and other gentlemen from Worcester, who got upon the train and went with it to Hopkinton, where it arrived at half past three o'clock and was received with a salute of artillery, the cheers of the populace and an address from the village authorities; after which, under escort of a company of riflemen, the whole party went to Captain Stone's tavern, where a collation had been provided. "While the party were at table the ladies were invited to take seats in the cars, and the military with their band of music to take a stand upon the tops of the cars, where they were formed in sections. In this manner they made an excursion of several miles down the road and back, which they appeared to enjoy highly. As they returned, the military on the tops of the cars approached the hotel with arms presented and music playing." Then followed speeches from Governor Davis and Governor Lincoln, and presently, at quarter after five, the party resumed their seats in the train and safely returned to Boston. The next day the train service was extended to Hopkinton.

A similar party which left Boston on the 15th of November in honor of the completion of the road as far as Westborough were not so fortunate as the Hopkinton excursionists. They started at eleven o'clock, but were delayed for some time at Needham from a cause which reads strange enough now—head winds. The road had but one track, and it was arranged that the excursion train should at Needham meet and pass another train on its way to Boston. It was autumn, and there was a gale blowing from the east, so that the locomotive *Meteor* had been delayed in its up passage " by a strong head wind," and consequently could not return on time. After a due amount of waiting the excursionists, however, lost their patience and determined to proceed. They accordingly did so; but when they had cautiously crept along for about four miles they met the belated *Meteor* coming down upon them before the wind. There was nothing for it but to take the back track to Needham, and there get out of its way. This was done, and in doing it so much time was consumed that the train did not reach Westborough until two o'clock,—having accomplished thirty-two miles in the space of three hours. Once at Westborough, however, the party proceeded to celebrate with a dinner and speeches after the usual fashion. The local authorities welcomed the directors with an address, to which Mr. Hale replied. At the close of his remarks he ventured to say, in reference to the railroad, that, " some of them hoped to see this work of improvement extended much further —to behold the day when the city of Boston would be placed within an afternoon's ride of the rich valley of the Connecticut River, and even when the banks of the Hudson would be brought within a day's journey of the metropolis of the commonwealth." He concluded by expressing a wish that they might all live to see these anticipations realized. He himself certainly did live to see them realized; for the road was formally opened from Boston to Albany in December, 1841, just seven years from the time at which he spoke, and before his death in 1863 the question was already, not of connecting Boston with Albany, but the seaboard of the Atlantic with that of the Pacific.

Among the other speakers at Westborough were Gov. Davis and both of the Everetts, Alexander and Edward. A Mr. P. P. F. De Grand was also there, a French emigrant of the old type, long resident in Boston, whose name was very closely connected with the early history of Massachusetts railroad development. Indeed Mr. De Grand was at times almost the moving spirit of the first enterprises, and it was his wont energetically to refer to the Worcester railroad as being a forty-four mile extension of Boston Long Wharf. During the periods of discouragement which, a few years later, marked certain stages of the construction of the Western road, connecting Worcester with Albany,—when both money and courage seemed almost exhausted,—Mr. De Grand never for a moment faltered. He might almost be said to have then had Western railroad on the brain. Among other things, he issued a certain circular which caused much amusement and not improbably some scandal among the more precise. The Rev. S. K. Lothrop, then a young man, had preached a sermon in Brattle Street church, which attracted a good deal of attention, on the subject of the moral and Christianizing influence of railroads. Mr. De Grand thought he saw his occasion, and he certainly availed himself of it. He at once had a circular printed, a copy of which he sent to every clergyman in Massachusetts, suggesting the propriety of a discourse on " the moral and Christianizing influence of railroads in general, and of the Western railroad in particular."

To return, however, to the Boston & Worcester road. It was completed in June, 1835, just four years after it had been chartered. It ran through a far from difficult country and was but forty-four miles in length; but in those days eleven miles a year were looked upon as quite a rapid rate of railroad construction. At last, on the 3d of July, a locomotive with one passenger car, in which were a few of the directors, passed over the line from end to end, and on Saturday, the 4th, the four engines, which constituted the entire motive power of the company, passed twice each way between Boston and Worcester,—two engines, drawing eleven cars, leaving the opposite ends of the road at the same time. During the day over fifteen hundred passengers were carried. In reality the opening of those first completed railroad lines,—for the Worcester was the third of the original lines all opened within one month,—was an event to be celebrated,—an event second in importance to none in the history of the State. The formal opening took place two days later, on the 6th, which, as hardly needs to be said, was a great day for Worcester, then a quiet country town of six thousand inhabitants. At ten o'clock in the morning a special train, consisting of twelve cars, drawn by two locomotives, and carrying some three hundred officials and invited guests, left Boston and reached Worcester at about one, where its arrival was welcomed by a salute of artillery and a general ringing of bells. Charles Allen, afterwards a member of Congress and Chief

Justice of the Superior Court of the State, was chairman of the committee of arrangements, and the programme included a procession and dinner at the Town-Hall. This, as usual, was followed by a great many toasts and a considerable effusion of after-dinner eloquence. Mr. Hale spoke, of course, as president of the road. He was followed by Edward Everett, as Governor of the Commonwealth; who, the report says, made a speech "of uncommon beauty and interest, interspersed with pleasant anecdote and eloquent remark. It was one of the happiest efforts of the distinguished gentleman."

Of greater interest to posterity, however, were the more common place remarks, from a rhetorical point of view, of Mr. Henry Williams of Boston, one of the directors as well as the clerk of the company. He took occasion to allude "with much feeling to the difficulties with which the enterprise had to contend at the outset without the aid of the capitalists, who hesitated to embark in so perilous an adventure. 'The work was commenced and has been completed,' said Mr. Williams, 'by the middling class in the community.'" For the rest the toasts on this occasion were of much the usual character, and now, after the lapse of more than forty years, they have certainly lost what little of flavor they may perchance once have had, and read thin and flat enough. One only among them has a curious sound, being strongly suggestive of the familiar "tempora mutantur" of the Latin poet. It was offered by ex-Governor Davis and was in these words,—"Railroads. We are willing to be rode hard by such monopolies." On that day certainly no anticipation of Granger agitations or of "Potter" laws was present in the mind of Governor Davis, or of any one else. At least no such forebodings troubled the festivities of the occasion, and a prophet who would then have dared to predict that within the lifetime of any there gathered together a political party would rise up with whom the indiscriminate denunciation of those who built railroads as "vampires" and those who operated railroads as "the robber barons of modern civilization" was the breath of life,—assuredly that prophet of evil would have found himself compelled to face a storm of jeers and contumely. But no such bird of ill-omen presented himself, and at eight o'clock that evening the safe and uneventful return home of the Boston party brought the celebration to a close.

Meanwhile those who had the construction of the other roads in charge were not idle. The Boston & Providence was fast approaching completion, and on Tuesday, June 2d, a party went over it from end to end. The following account of the excursion, which appeared in a Providence paper a few days later, has rather acquired than lost interest through the lapse of forty years. The party set out from Providence upon the invitation of the board of directors:

"It was in contemplation to have taken the new engine that had arrived from Philadelphia only the day before, but some of her pipes were not in order, and we finally set off from the depot at India Point, at a quarter before one o'clock in the afternoon, with two cars, each propelled by *two horse power*.

"The application of horses afforded us a most fortunate opportunity for inspecting the grand structure over which we passed. The road . . has been laid to endure with the everlasting hills, and is finished with a neatness very gratifying to the eye. The viaduct at Canton, though yet unfinished, is a stupendous work. A view of it many times repays the trouble of passing round. The excavations and embankments in Canton are also worthy of minute attention; they testify in strong language, to man's dominion over nature, and his ability to overcome any obstacle to any undertaking that is not either morally or physically absurd. The project of cutting through these rocky heights and crossing the valley of the river by the viaduct was a very bold one. A hesitating mind would have surmounted them by stationary engines, or some less formidable way. But any other mode would have detracted very much from the facilities which give value to such a road.

"The road has been constructed under the direction of Major McNeil, and it will stand for ages, an enduring monument of the high talents and high attainments of its accomplished engineer.

"Among the curiosities on the way is a bog in Mansfield where the road sunk, during its formation, to the depth of forty feet; and it is also a curious fact that sixteen miles and a half of this road are on a perfectly straight line.

"After examining the work at Canton we took the engine at twenty minutes past five, and were landed at West Boston at about six o'clock. The party accompanied the directors to their *depot* at the Tremont House, and enjoyed their overflowing hospitality with keen appetites and grateful hearts."

The viaduct at Canton, by the way, the bold conception and fine construction of which excited so much admiration in the minds of these excursionists, really was a most creditable piece of work. When they did build, they built better in those days than they now do, and the passage of forty years of constant use has developed no greater need for repairs on the Canton viaduct than it has on the pyramids of Egypt. That viaduct, also, has a history of its own, curiously illustrating the value of fatal accidents as a dynamic force in railroad development. In the earliest surveys of the route of the Boston & Providence line, the deep and wide ravine through which the Neponset river flows between the elevated hills of Canton had presented itself as a serious obstacle in the way of the enterprise. Closely following the precedents already established in the case of the coal tramways, both at home and abroad, the engineers at first proposed to overcome the difficulty by means of inclined planes operated by stationary engines. It so happened, however, that there already was, only a few miles away and also close to the banks of the Neponset, an inclined plane at the end of the Quincy Granite Railway. From motives of curiosity parties of gentlemen were in the custom of visiting that work. While the construction of the Boston & Providence road was yet undecided, a party of this sort was one day ascending the Quincy incline when the pulleys broke, and the car on which the visitors were ran backwards down the grade, throwing them out and killing one of them, a Mr. Gibson, well known in Boston. This accident brought inclined planes into great disfavor, and induced the construction of the

Canton viaduct; just as invariably since that time every serious railroad accident has had a direct and often most perceptible influence in bringing about some great advance in railroad construction or appliances for safety.

To return, however, to the party of excursionists who visited Boston in the early days of June, 1835. They had left Providence on Tuesday. On the very day before, another event of interest had taken place on Long Island sound, for the ill-fated steamer *Lexington*, built specially to run between New York and Providence in connection with the new railroad, had then made her trial trip. The *Lexington* was constructed for and under the direction of Captain Cornelius Vanderbilt, who had not then arrived at his subsequent title by courtesy of Commodore, but " whose reputation for fast boats is,"—the contemporaneous authority goes on to say,—" so well established in this community." The *Lexington* upon her trial trip astonished those on board of her, and the editor of the New York *Courier and Enquirer*, who was one of the number, thus gave vent to his feelings:

"We were of the party who accompanied her on this novel and interesting expedition: and although the Boston & Providence Railroad is not yet opened—which event will shorten the time of travelling between those cities two hours—we yesterday, *breakfasted at Boston, left there at* 2 A. M., *and arrived in this city off Dry Dock in eleven hours and fifty-nine minutes* from Providence,—performing the entire distance in less than sixteen hours, and bringing with us the Boston daily papers of yesterday morning for the benefit of our readers and those of our cotemporaries."

In other words, General Webb had left Boston at two in the morning and arrived in New York at six o'clock on the evening of the same day, being the shortest time which had ever been made between those two cities. He then proceeds, in a strain of enthusiastic exultation over the prospect of " reducing the time of overcoming the distance between New York and Boston (250 miles) to fourteen hours," and closes with a tribute which, though offered nearly half a century ago, still has an amusing significance :

" Other sections of the country will be equally benefited by this improvement of steam navigation by Captain Vanderbilt, and his name will in future be classed with those of Fulton and Stephenson, to the latter of whom we owe nearly all the improvements which have been made in the steam engine, since the death of that great man to whom the world is indebted for that most important discovery which has ever been made except the art of printing."

Having given one side of the picture; it is but fair to present the other. The advent of railroad locomotion was not even in America hailed by all in a similar spirit of exuberant satisfaction. A little over a month after the time when General Webb went from Boston to New York in sixteen hours, a gentleman of the very old school, then in his sixty-fourth year, made the same trip; and in his diary thus freshly recorded his experience and sensations:

"July 22, 1835.—This morning at nine o'clock I took passage in a railroad car (from Boston) for Providence. Five or six other cars were attached to the locomotive, and uglier boxes I do not wish to travel in. They were made to stow away some thirty human beings, who sit cheek by jowl as best they can. Two poor fellows, who were not much in the habit of making their toilet, squeezed me into a corner, while the hot sun drew from their garments a villainous compound of smells made up of salt fish, tar and molasses. By and by, just twelve,—only twelve —bouncing factory girls were introduced, who were going on a party of pleasure to Newport. "Make room for the ladies!" bawled out the superintendent. "Come, gentlemen, jump up on the top; plenty of room there." "I'm afraid of the bridge knocking my brains out," said a passenger. Some made one excuse and some another. For my part, I flatly told him that since I had belonged to the corps of Silver Grays I had lost my gallantry, and did not intend to move. The whole twelve were, however, introduced, and soon made themselves at home, sucking lemons and eating green apples. . . . The rich and the poor, the educated and the ignorant, the polite and the vulgar, all herd together in this modern improvement in travelling. The consequence is a complete amalgamation. Master and servant sleep heads and points on the cabin floor of the steamer, feed at the same table, sit in each other's laps, as it were, in the cars; and all this for the sake of doing very uncomfortably in two days what would be done delightfully in eight or ten. Shall we be much longer kept by this toilsome fashion of hurrying, hurrying, from starting (those who can afford it) on a journey with our own horses, and moving slowly, surely and profitably through the country, with the power of enjoying its beauty and be the means of creating good inns. Undoubtedly, a line of post-horses and post-chaises would long ago have been established along our great roads had not steam monopolized everything. . . . Talk of ladies on board a steamboat or in a railroad car. There are none. I never feel like a gentleman there, and I cannot perceive a semblance of gentility in any one who makes part of the travelling mob. When I see women whom, in their drawing-rooms or elsewhere, I have been accustomed to respect and treat with every suitable deference,—when I see them, I say, elbowing their way through a crowd of dirty emigrants or low-bred homespun fellows in petticoats or breeches in our country, in order to reach a table spread for a hundred or more, I lose sight of their pretentions to gentility and view them as belonging to the plebeian herd. To restore herself to her caste, let a lady move in select company at five miles an hour, and take her meals in comfort at a good inn, where she may dine decently. . . . . After all, the old-fashioned way of five or six miles with liberty to dine decently in a decent inn and be master of one's movements, with the delight of seeing the country and getting along rationally, is the mode to which I cling, and which will be adopted again by the generations of after times."*

Curiously enough, but probably as the result of a very natural spirit of emulation in those engaged in building them, the three initial roads of Massachusetts,—the germs of her subsequent railroad system,—were all completed and opened to traffic within four weeks of each other,—the Providence on the 11th of June, the Lowell on the 27th of the same month, and the Worcester on the 3d of July. They were all well built roads, especially that to Lowell, in the construction of which the Manchester & Liverpool precedents had been so closely followed that the serious error was committed of laying the rails on stone blocks instead of wooden ties. It is, indeed, matter of curious observation that almost uniformly those early railroad builders

* Recollections of Samuel Breck, pp. 275–7.

made grave blunders, whenever they tried to do their work peculiarly well; they almost invariably had afterwards to undo it. The Lowell road, for instance, was too well built in many respects. On a portion of its track the stone blocks, into which the oaken plugs to spike the rails to were inserted, were laid on a foundation of continuous, parallel, dry, stone walls running in trenches under each line of rails, and from two and a half to four feet deep and a foot and a half wide. Such work as this was intended to last, and doubtless the Boston & Lowell directors thought that they had acquitted themselves of their trust with a far seeing economy. Unfortunately, as time passed, experience decided the other way. They gradually learned to their great surprise that speed without elasticity is always costly; and to-day the sides of the roadbed are liberally ornamented with those useless stone sleepers, the eternal life of which was once looked forward to with confident pride. It was only through the shrewd sense of its constructing engineer, also, that the Boston & Providence company was saved from this same blunder. Captain McNeill had been sent abroad to examine the Manchester & Liverpool road. While doing so he not only had the sense to see that the objections to the use of wood which existed in England, because of its scarcity, did not hold good in this country, but he also with great sagacity divined at once the importance of an elastic road-bed. In one important respect, however, the early railroad companies enjoyed an enormous advantage over those of to-day; the materials they used were as a rule honestly made. The original iron of the Boston & Providence weighed fifty-five pounds to the yard and would outlast most modern steel. The last of it was not taken out of the tracks until 1860, and then, after twenty-five years of continuous service, it was still in good condition.

The first epoch of railroad construction in Massachusetts did not, however, close with the opening of the Boston & Worcester road. On the contrary, it rather began with that want. It closed six years later —during the last days of 1841—when at length with hard struggle and after many and bitter discouragements, at times verging almost on despair, the Western railroad was completed. By it Boston was brought into a close connection with Albany and that great network of internal communication, whether by land or water, which there found an outlet. For the time the construction of that road was really a great achievement,—much greater than the subsequent building of the Pacific railways. Begun in 1834, it was seven years, covering all the dreary period which followed the panic of 1837, before it could be finished. During that time the work progressed at an average rate of about twenty two miles a year. Repeatedly it would have come to a dead stand-still had not the assistance of the State been extended to it with a liberal hand. Of its original officers during the period of construction but one now survives, Josiah Quincy, then the younger of the name, who was treasurer of the corporation. As in the case of Mr. De Grand, the sanguine temper of Mr. Quincy was then of no little service. During the too frequently recurring days of despondency he was wont to humorously draw courage for himself and his associates from the remark of that king of Spain who met the suggestion of a canal between two points in his dominions with the dry negative, that "If the Almighty had intended there should be navigation between these two points, he would undoubtedly have placed a river there; but it was not for a poor mortal like himself to improve on the infinite wisdom of God's handiwork." But in the case of the Berkshire hills, as Mr. Quincy argued, the Almighty had made a practicable roadway, and hence it was clear He meant in His wisdom there should be a railroad built through them, and consequently the road would be built. Built at last it was, and its completion brought to a triumphant close the first epoch of Massachusetts railroad construction. The State then had a complete railroad system; and, in closely studying the records of the time, it is curious to see what a revolution the new power had already brought about. The community had in 1841 fully entered on the new life. Accordingly when the Western railroad was at last opened, though the event was one of too much importance to be passed over unnoticed, the celebration had distinctly lost that fresh, primitive flavor which alone lends to our times a charm in the earlier occasions. The opening of a new railroad was in 1841 an old story. Every one had then made journeys by rail. There was no longer any novelty about the thing, and the orators who tried to excite fresh emotions of wonder by dwelling on the well-worn theme began to find it very hard work. Nevertheless on the 27th of December 1841, the members of the Boston city government started for Albany, on what would now be termed a " municipal junket." Among other invited guests they took with them a delegation from New Bedford. To these New Bedford gentlemen was due on this occasion the last vestige of that simple wonder which had always been so prominent a feature in the earlier railroad celebrations. In order to lend point to the astonishing fact that, leaving their homes in the morning they would in fifteen hours be in Albany, these gentlemen during the small hours of the day of their departure caused some spermaceti candles to be moulded. These they took with them on their trip, and that evening the rays from these candles illumined the table around which took place the civic banquet at Albany. But the Albanians were not to be outdone. They were to return to Boston with their guests the next day; and in doing so, they took with them a barrel of flour, the wheat for which had been threshed at Rochester on the previous Monday,— they went to Boston on Wednesday—while the barrel itself was made from wood which on the threshing day had been growing in the tree. This flour, duly converted into bread, the authorities of the two cities and their invited guests solemnly ate at a grand dinner given at the United States Hotel in Boston on the evening of December 30th, 1841. Of the toasts and speeches given utterance to on this occasion there is little enough to say. In them honest astonishment had given place to a mouthing eloquence. Every one realized fully the importance and the far reaching consequence of the event they were met to celebrate,—the fire companies and the military were all paraded and the air was filled with the strains of music,—but none the less it was all a twice-told tale. Railroads had grown to be commonplace affairs. The world had already accustomed itself to the new conditions of its existence, and wholly refused to gape in childish wonder at the thought of having accomplished a journey of fifty miles more or less between the rising and setting of even a December sun. The genesis of the system was complete.

THE RAILROAD PROBLEM.

DURING the last ten years there has been so much vague discussion of what is commonly known as the Railroad Problem, that many people, and those by no means the least sensible, have begun gravely to doubt whether after all it is not a mere cant phrase, and whether any such problem does indeed exist. Certainly the discussion has not been remarkable for intelligence, and the currency question itself has hardly been more completely befogged in clouds of indifferent declamation, poor philosophy and worse logic. No fallacy has been too thin to pass current in it; and the absolute power which certain words and phrases have held over the public mind has throughout seemed to set both argument and patience at defiance. Under these circumstances, before beginning to discuss the Railroad Problem, it might seem proper to offer some definition of what that problem is. To do this concisely is very difficult. As an innovating force the railroad has made itself felt and produced its problems in every department of civilized life. So has the steam-engine; so has the newspaper; so has gunpowder. Unlike all these, however, the railroad has developed one distinctive problem, and a problem which actively presses for solution. It has done so for the reason that it has not only usurped, in modern communities, the more important functions of the highway, but those who own it have also undertaken to do the work which was formerly done on the highway. Moreover, as events have developed themselves, it has become apparent that the recognized laws of trade operate but imperfectly at best in regulating the use made of these modern thoroughfares by those who thus both own and monopolize them. Consequently the political governments of the various countries have been called upon in some way to make good through legislation the deficiencies thus revealed in the working of the natural laws. This is the Railroad Problem. Thus stated, it hardly needs to be said that the questions involved in its solution are of great magnitude and extreme delicacy. To deal correctly with them requires a thorough knowledge of intricate economical laws, superadded to a very keen insight into political habits and modes of thought. For not only is there a general railroad problem for all countries, but this problem has to be dealt with in a peculiar way in each country. One mode of treatment will not do for all. Before discussing, therefore, the form this problem has assumed in America it will be well to briefly review its development, and the efforts made to solve it elsewhere. The experience of other countries can hardly fail to throw a side light at least on the direction events are taking here.

The railroad originated in England, and in England it has upon the whole attained its highest present stage of development. The English railroad system and the English experience must, therefore, first be described. In one of the earlier parliamentary debates on the subject of railroads the Duke of Wellington is reported to have said that in dealing with them it was above all else necessary to bear in mind the analogy of the king's highway. The remark was certainly characteristic, both of the individual and the race. Without any careful analysis to find out whether it was real or apparent only, the analogy was accepted and upon it was based that whole elaborate system of legislation through and in spite of which both in Great Britain and in America the railroad system grew up, and in the meshes of which it is now struggling. In fact the analogy was essentially a false one. In no respect did the railroad in reality resemble the highway, any more than the corporation which owned and operated it resembled the common carrier. The new system was not amenable to the same natural laws which regulated and controlled the operations of the old one, and the more the principles and rules of law which had grown out of the old system were applied to it, the worse the result became. The acme of the ludicrous in this respect was, however, reached not in England but in America. In England the truth dawned in time on the minds of those upon whom the work of legislation devolved. After more than forty years of blundering it was there at last realized in 1872 that the railroad system was a thing *sui generis*,—a vast and intricate formative influence, as well as a material power, the growth of which was to be curiously watched in the expectation that in due time it would develop some phase which again would call forth a corresponding development in the machinery of government, through which its political and economical relations with the community would be finally established on some rational and permanent basis. Meanwhile at the very time this result was reached in Great Britain, and the railroad problem consequently ceased to be a matter for active discussion, America was clinging more desperately than ever to that false analogy which had thus been finally abandoned in the place where it originated. Since 1872, even more than before that time, the American legislation has been inspired by the theory that the railroad corporation is nothing but an overgrown common-carrier, who has in some way got the monopoly of a highway, and, being crazed by sudden and ill-gotten gains, has forgotten his proper place in life; of which he must forthwith be reminded through an exercise of political power. The old analogy suggested by the Duke of Wellington, as mischievous as it is false, still maintains a strong hold on the legislative mind and belittles a great question.

Upon it, however, the whole railroad system of Great Britain was founded. In the first place, the proprietor of the road-bed and the carrier over it were to be different persons. Provision in this respect was especially made in all early charters, and it was supposed that the power of using the road, which was reserved to all the world on certain fixed terms, would make impossible any monopoly of the business over it. Experience, of course, quickly showed how utterly fallacious this reasoning was. No glimmer of doubt, however, as to the correctness of the analogy drawn from the king's highway suggested itself to the parliamentary mind. On the contrary it was only the more tenaciously clung to. Recourse was had to a system of fixed maxima charges, and the old tollboards of the turnpikes were incorporated at enormous length into the new charters as they were granted. One of these, for instance, which went through Parliament in 1844, consisted of three hundred and eighty-one distinct sections, in which, among other things, it was prescribed that for the carriage of a "horse, mule, or ass" the company might charge at a rate not to exceed three pence per mile, while for a calf or a pig or "other small animal," the limit was a penny. Naturally, this attempt at regulation proved no more efficacious than the other; but it served its turn until yet another theory, that of parallel highways controlled by competing common carriers, was ready to be developed. This was about the year 1840. The chaotic condition of the English railroad legislation had then begun to attract public notice, and this led to a reference of the whole subject to the first of those many special parliamentary committees which have taken it into consideration. Sir Robert Peel was a member of this committee, which apparently fell back on the principles of free trade as affording all the regulation of railroads which was needed.

It was argued that "an enlightened view of their own interests would always compel managers of railroads to have due regard to the general advantage of the public." At the same time, to afford railroad managers a realizing sense of what the principles of free trade were, numerous charters were granted and liberal encouragement given to the construction of competing lines. Then came on the great railroad mania of 1844, and, as other countries have since done, England awoke one day from dreams of boundless wealth to the reality of general ruin. Free trade in railroads was then pronounced a failure, and in due time another parliamentary committee was appointed, and the whole subject was again taken into consideration. Of this committee Mr. Gladstone was the guiding spirit. Meanwhile Sir Robert Peel, who was then prime minister, had changed his mind as respects the efficacy of "an enlightened self-interest" stimulated by competition, and had come to the conclusion that railroad competition was an expensive luxury for the people indulging in it, and that there might be something in state management of railroads; a system which his friend, King Leopold of Belgium, was then developing with much judgment and success. Accordingly Mr. Gladstone's committee made a series of reports which resulted in the passage of a law looking to the possible acquisition of the railroads by the state at the expiration of twenty-one years from that time. With this measure as the grand result of their labors the committee rested. Not so the railroad system. The twenty-one years elapsed in 1865, and during that time Parliament sat and pondered the ever-increasing complication of the railroad problem with most unsatisfactory results. Competition between railroads through all those years was working itself out into combination; and, as the companies one after another asked and secured acts of amalgamation, obstinately refusing to compete, it was clearly perceived that something was wrong. The parliamentary mind was sorely troubled; but no way of deliverance revealed itself. In 1865 a new commission was appointed, which went again over the familiar path, this time in the direction of state ownership. The cry now was that the process of amalgamation, or consolidation as we in America term it, had gone so far that the time was close at hand when the railroads would manage the state, if the state did not manage the railroads. In truth there was something rather alarming in the speed with which illustrations followed one upon another of the truth of George Stephenson's aphorism, that—"Where combination is possible, competition is impossible." The thing, too, was now done upon a scale of magnificence which was not less startling than novel. The world had seen nothing of the kind before, and naturally paused to ask what it all meant and whither it was tending. For instance, one committee pointed out, as an example of what the process might lead to, that a single amalgamation was suggested to it through which a union of 1200 miles of railroad would be effected, bringing under one control £60,000,000 of capital with £4,000,000 of annual revenue, and rendering impossible throughout one large district the existence of an independent line of railway. A few years later, when the next committee sat, all this had become an established fact; only the mileage was 1500 instead of 1200, the capital £63,000,000 instead of £60,000,000, and the annual income £7,000,000 instead of £4,000,000. Nevertheless the commission of 1865 followed closely in the steps of its predecessors. It dumped upon the tables of Parliament an enormous "blue-book," which left the matter exactly as dark as it was before. Still the amalgamations went on. All England was rapidly and obviously being partitioned out among some half-dozen great corporations, each supreme in its own territory. Then at last, in 1872, a committee on railroad amalgamations was appointed, the Marquis of Salisbury and the Earl of Derby being two of its members, which really gave to the whole subject an intelligent consideration. Unlike its predecessors, that committee did not leave the railroad problem where they found it. On the contrary, they advanced it by one entire stage on the road to its solution. In the first place, after taking a vast amount of evidence, they proceeded to review the forty years of experience. The result of that review may be stated in few words.

They showed with grim precision how, during that period, the English railroad legislation had never accomplished anything which it sought to bring about, nor prevented anything which it sought to hinder. The cost to the companies of this useless mass of enactments had been enormous, amounting to some £80,000,000; for these were 3,300 in number and filled whole volumes. Then the committee examined in detail the various parliamentary theories which had, at different stages, marked the development of the railroad system. The highway analogy was dismissed in silence; but of the "enlightened view of self-interest" theory, it was remarked that experience had shown that as a regulating force this was to be relied upon "only to a limited extent." The principle of competition was next discussed, and the conclusion of the committee was "that competition between railroads exists only to a limited extent, and cannot be maintained by legislation." Of the great Gladstone act of 1845, looking to the ultimate purchase of the railroads by the government, it was remarked that "the terms of that act do not appear to be suited to the present condition of railway property, or to be likely to be adopted by Parliament, in case of any intention of Parliament at any future time to purchase the railways." Having disposed of this measure, the committee addressed itself to the amalgamation panic, which through so many years had rested like a nightmare on the slumberous discussions of Parliament. They cited the case of the North-Eastern railway, which was composed of thirty-seven once independent lines, several of which had formerly competed with each other. Prior to their consolidation these lines had, generally speaking, charged high rates, and they had been able to pay but small dividends. Now, the North-Eastern was the most complete monopoly in the United Kingdom. From the Tyne to the Humber it held the whole country to itself, and it charged the lowest rates and paid the highest dividends of all the great English companies. It was not vexed by litigation, and while numerous complaints were heard from Lancashire and Yorkshire, where railway competition existed, no one had appeared before the committee to prefer any complaint against the North-Eastern. In view of such facts as these the committee reported that amalgamation had "not brought with it the evils that were anticipated, but that in any event long and varied experience had fully demonstrated the fact that while Parliament might hinder and thwart, it could not prevent it, and it was equally powerless to lay down any general rules determining its limits or character." The statute-book was full of acts regulating the rates at which the poorer classes should be carried by rail, and these acts at least had always been pointed to as indisputable evidence of the virtue and efficacy of railroad regulation by Parliament. In their day they had perhaps done good service; but yet even of these as a whole it was reported that "the ill success of this attempt may well justify hesitation in entering upon further general legislation of the same kind." Finally, the committee examined all the various panaceas for railroad abuses which are so regularly each year brought forward as novelties in the legislatures of this country. These they passed in merciless review.

Equal mileage rates they found inexpedient as well as impossible; the favorite idea of a revision of rates and fares with a view to establishing a legal tariff sufficient to afford a fair return and no more on the actual cost of the railroads, they pronounced utterly impracticable; tariffs of maxima charges incorporated into laws, they truly said had been repeatedly enacted and as often had failed; periodical revisions of all rates and fares by government agents they found to be practically impossible, unless some standard of revision which had not yet been suggested could be devised. There is in the French law a provision that whenever the profits of any road shall exceed a certain percentage on its cost, such excess shall be divided between the corporation owning the railway and the government. This plan, also, the committee took into careful consideration, only to conclude that in Great Britain its adoption would be attended with " great if not insuperable difficulties." Finally, the owning of the railroads by the government was referred to as " a state of things which may possibly arise," but one which the committee was not at all disposed at present to recommend.

At first glance, therefore, it seemed as if this committee had arrived at only negative results; but in truth they had reached positive conclusions of the first importance. They had, indeed, clearly stated the problem; a thing never before done in Great Britain. The natural development of the railroad system as a system was recognized, and the folly of restrictive legislation demonstrated. A new policy was thus established, at the base of which was the principle of private ownership and management, which was to be left to work out its own destiny through that process of combination in which competing monopolies always result. The members of the committee saw perfectly clearly where their process of reasoning would bring them out. It could result only in a tacit assent to the growth of private corporations until they become so great that they must, soon or late, assume relations to the government corresponding with the public nature of their functions. This was obvious enough. Meanwhile the committee also saw with equal clearness that this was a question of the future,—perhaps of the remote future; a question which certainly had not yet presented itself, and which they had no disposition to precipitate. They accordingly fixed definitely the policy of Great Britain as an expectant one. The railroad system was to be left to develop itself in its own way, as a recognized monopoly, held to a strict public accountability as such. Whenever it should appear that it abused its privileges and power, then the time for action would have arrived. As yet this was not the case in any such degree as called for a decisive and far reaching measure of reform. To better watch over it meanwhile, and to cause its members to work with less friction among themselves and with a more careful regard to the equal rights of private persons, an exceptional piece of governmental machinery was provided in the form of a board of commissioners. The powers of this tribunal were both judicial and executive in character and very broad. It was its duty not only to hear all complaints of private parties, but to intervene in cases of difficulty which might arise between the companies themselves. The board was in fact designed to insure to the community an easy and equitable interchange of traffic over its railroad lines, as well as to put a stop in so far as might be practicable to that unjust and vexatious system of discrimination which seems to be inseparable from all active railroad competition. Through this board the exceptional character of the railroad system was at last recognized. An attempt was made to deal with the anomaly as an anomaly. Meanwhile the most noticeable feature in the career of the board, which has now been in existence five years, is the very trifling call which seems to have been made upon it. So far as can be judged from its annual reports, the cases which come before it are neither numerous nor of great importance. It would, however, be wholly unsafe to conclude from this fact that such a tribunal is unnecessary. On the contrary, it may confidently be asserted that no competent board of railroad commissioners, clothed with the peculiar powers of the English board, will ever, either there or anywhere else, have many cases to dispose of. The mere fact that the tribunal is there, —that a machinery does exist for the prompt and final decision of that class of questions, puts an end to them. They no longer arise. They cease to arise for the simple reason that the railroad corporations in these matters are not like the members of a numerous and complicated business community. The controversies among them, which do not involve legal points, are comparatively simple and confined to few persons. These they will in the vast majority of cases settle themselves, if they only know that if they do not so settle them, a public official will. The English board has always been composed of very competent .men. Accordingly the officials of the corporations, knowing quite well in advance what their decisions will be, do not probably care to encounter them. A single test case disposes of innumerable subsequent cases.

In Great Britain, therefore, the discussion of the railroad problem may be considered as over for the time being. It is quiescent, not dead. The period of meddlesome and restrictive legislation is passed, and the corporations are now left to work out their own destinies in their own way, just so long as they show a reasonable regard for the requirements and rights of the community. The time may not be remote, when, for instance, all England will be served by three or four gigantic railroad companies, or perhaps by only one; just as many cities are now furnished with gas. Nor is this ultimate result any longer viewed with apprehension. The clearer political observers have come to realize at last that concentration brings with it an increased sense of responsibility. The larger the railroad corporation, the more cautious is its policy. As a result, therefore, of forty years of experiment and agitation, Great Britain has on this head come back very nearly to its point of commencement. It has settled down on the doctrine of *laissez faire*. The river is not to be crossed until it is reached; and, perhaps, by the time it is reached a practicable method of crossing will have become quite apparent.

Turning now from Great Britain to Belgium, an opportunity is offered to observe the practical working of a wholly different policy. The famous Belgian railroad system originated with King Leopold, and bears to this day marks of the creating mind. When the Manchester & Liverpool railway was completed the Belgian revolution had not yet taken place, and Leopold was still a resident of England. His attention was naturally drawn to the possible consequences of this new application of steam, and when, a few years later, he was called to the throne of Belgium, one of his earliest projects related to the construction of railroads in his new dominions. He was strongly persuaded, however, that the English system of private construction was not the correct one. He, as well as the Duke of Wellington, strongly adhered to the analogy of the highway; but, more logical than the duke, his was the king's highway and not a turnpike. Accordingly he planned a system of railway communication in which the roads—the steam-highways—were to be constructed, owned, and operated by the state. With some difficulty legislative assent to his scheme was obtained, and the earliest lines were undertaken in 1833. The government then went on year by year developing the system, but failed to keep pace with the public demand. Accordingly, in a few years, though not until after the principal and more remunerative routes were occupied, concessions, as they were called, being the equivalent of English charters, were made to private companies, which carried on the work of extension.

One peculiar feature in all these concessions had, however, a direct and sagacious though somewhat distant bearing on the fundamental idea of the Belgian railroad system,—that of ultimate government ownership. They were all made for a term of ninety years, at the expiration of which the railways were to become the property of the state, which was to pay only for their rolling-stock. The right was also reserved to the government of buying back the concession at any time, upon assuming payment to the owners for any unexpired balance of the ninety years of a yearly sum equal to the average net receipts during the seven years next preceding the taking.

Until their concessions should be thus terminated, however, the private companies owned and operated their several roads in much the same way as English or American corporations; although the greatest benefit from their construction resulted to the state lines, which, holding the centre of the country and the main routes of communication, kept the private lines necessarily tributary. In 1850, the government owned about two-thirds of all the railroad mileage then in operation, and private companies the other one-third. Ten years later the proportion had changed, two-thirds of the system being in the hands of private companies. It so happened, also, that, as the government in making the concessions had followed no plan of districting the country, but had rather adopted a policy of competing lines, these lines competed not only with each other but also with the state lines. From this fact there resulted a condition of affairs which was wholly unanticipated, but which has since constituted the very essence of the Belgian railroad system. For the first and only time in railroad history, a case was presented in which competition did not result in combination. The one system of lines being owned by the state and the other by private companies, no consolidation of the two was practicable as against the public; and accordingly the government found itself in a position to regulate the whole system through the ownership of a part of it, and in consequence was able to establish a policy of cheap railroad transportation, under the influence of which the country developed with amazing rapidity.

The action of the government, however, practically forced the various independent companies to unite among themselves; until, about the year 1860, they had become consolidated into trunk lines sufficiently powerful to compete with the state on equal terms. Under these circumstances, in order to maintain the principle of its railroad system, the government was forced into a policy of further development which in 1870 resulted in the acquisition by lease of a whole system of competing lines. Again, in the year 1872, as one of the indirect outcomes of the Franco-German war, the government felt constrained to purchase, from the English company which owned it, the Luxembourg road in order to keep it from falling into German hands. Finally, at the close of the year 1876, the state owned or controlled sixty per cent. of the entire railroad mileage of the country, while the remaining forty per cent. belonged to private companies. These private companies practically operate their roads with the utmost freedom from governmental interference. They raise and lower their rates at discretion, and no limitation is put on the amount of dividends they may declare. In respect to questions of police and safety only does the government formally interfere with them; and, with the exception of certain guaranteed lines, it has no power even of supervising their accounts, or, indeed, of compelling them to render any.

Of late years, therefore, Belgium has simply presented the spectacle of the state, in the character of the richest and most powerful railroad company of its system, holding in check and regulating other companies, not greatly inferior to it in power, which compete with it for business and with which it deals on terms of equality. The effect of this on each system of roads has been excellent. At times, when the government has attempted certain great measures of reform or bold experiments in transportation, its course has been vehemently criticised by the private companies, who have complained that their property was being unjustly depreciated by tariff reductions made upon unsound principles, but which, from their position, they were compelled to adopt. This was perfectly true; but, on the other hand, the government was so largely interested in railroad property that it felt no disposition to persist in any line of experiment which seemed likely to reduce its value permanently; and in the long run the private companies have found that the experiments of government were far less to be feared than the wild and ruinous fluctuations of railroad competition, as it was carried on in Great Britain. These they were exempt from. The competition they had to meet was decided, but of a wholly different character. It was certain, firm, and equably distributed. Those managing the state roads acted at all times under a heavy sense of responsibility; they did not dare to show preference to persons or localities; they could not do business for anything or nothing one day, and the next combine against the public to make good their losses through extortionate charges. In a word, it was found that while the competition between private roads in Great Britain and America disturbed and disorganized railroad traffic, that between public and private roads in Belgium regulated it.

The government, meanwhile, in its turn pressed by the competition of the private lines, found itself compelled to work its roads on regular "commercial principles." In order to get business it made special rates, and, if necessary, entered into joint-purse arrangements with its adversaries. It made bold experiments, and through those experiments established certain principles of transportation now universally recognized. At other times its experiments resulted in failure and were abandoned. Yet little doubt can be entertained that it was the constant pressure of competition which kept the state lines up to their work and in the advance of railroad development. The tendency in Belgium now is for the government to absorb all the remaining lines. Should this be done, it will then remain to be seen whether by so doing that equilibrium to which the success of the whole system has apparently been due will not have been destroyed. Competition, certainly, will then no longer exist, and with its disappearance a strong incentive to activity may also disappear.

It would of course be most unnatural to suppose that the state roads of Belgium have always given perfect satisfaction to the community. There have, on the contrary, been very grave and distinct complaints in regard to their management, but nothing which will compare with those constantly made both in Great Britain and in America. To satisfy every one always is a result not likely to be attained under any system or in any country; meanwhile, it may with tolerable safety be asserted that the Belgian system is as satisfactory to the people of Belgium as the nature of things human permits that it should be; certainly the public feeling points very distinctly towards the acquisition of the remaining lines of the system by the government, while the sale of the government lines to private corporations has never been urged by any considerable party. Financially the undertaking has proved a decided success, the government roads netting an annual profit of late years of about six per cent, which is equivalent to at least ten per cent in this country.

While in Great Britain, therefore, the railroad problem seems entering upon a period of comparative quiescence,—a phase of expectancy, as it were,—in Belgium the contrary would seem to be the case. Should the Belgian government now adopt a policy of expansion, and proceed to acquire the remaining lines of the system, it will enter upon the very doubtful experiment of exclusive state management. The problem will then assume wholly new, phases, the development of which will everywhere be watched with deep interest.

The railroad system of France was developed on

principles wholly different from those adopted in England and Belgium. In that country there was none of the bold English initiative ready to force the experiment along through private enterprise ; nor was there any King Leopold on the throne. There was already an admirable system of highways, and, comparatively speaking no great need was felt of railroads. Moreover, in spite of the political changes and the turbulence which have characterized the history of the country, the French mind is essentially conservative ; it looks naturally to the government for an initiative, and not only submits to, but craves minute regulation from a central authority. Accordingly, when forty years ago England and America caught eagerly at the idea of railroad development, and rushed into it with all the feverish ardor which ever marks private speculation, France hung back. It was not until 1837, when already what are now the great trunk routes of Great Britain and of America had assumed a definite shape, that the French system began slowly to struggle into life. Even then the first attempts resulted only in failure. The government, after hesitating long, recoiled from the idea of following the bold example set by Belgium, and decided in favor of a system of concessions to private companies, instead of construction by the state. These companies were organized at last, and an appeal was made to the public. The public, still timid and lacking confidence in itself, failed to respond, and the companies, frightened at the liabilities they had incurred, renounced their concessions. Then at last, but not until 1842, the government definitely took the lead. A division of risk was effected. Nine great lines were mapped out, seven of which were intended to connect Paris with the departments of the frontier or the sea-board, while two were provincial. As respected some of these the state assumed the expense of acquiring the necessary lands and building the stations, while the companies undertook to furnish the superstructure and material, and to operate the roads; as respected others the companies took upon themselves the whole burden. The political disturbances of 1848 and the years immediately ensuing greatly retarded French development in railroads, as it did in everything else. It was not until 1859 that the system assumed a definite shape. Then at last, under the inspiration of the imperial government, a new and final arrangement was effected. The existing lines were consolidated, and France was practically partitioned out among six great companies, to each of which a separate territory was allotted. The fundamental distinction between the French and the English and American railroad systems was now brought into sharp prominence. Not only was no provision made for competition, but every precaution was taken to prevent it. No company was to trench upon the territory allotted another, and, in consideration of this immunity, each line undertook within its own district a railroad development proportionate to all reasonable demands. Again, however, the companies found the burden they had assumed out of proportion to their resources. Once more recourse was had to the state. The necessary assistance was forthcoming, but on condition. The lines to be constructed and operated by each company were laid down, and arbitrarily divided into classes, designated as the *ancien réseau* and the *nouveau réseau*, the first of which included the older and more profitable, and the latter the additional routes the construction of which was deemed essential. Upon the securities issued to build these last the government guaranteed a minimum rate of interest, which the companies undertook ultimately to reimburse. The material of both the *ancien* and the *nouveau réseaux* was also pledged as security for any advances which the state might be called upon to make. The amount of advances thus made up to the present time somewhat exceeds $60,000,000. The concessions are for ninety-nine years, at the expiration of which the roads will revert to the state, which is bound, however, to purchase the rolling-stock at a valuation, after deducting advances made. The right is also reserved to the government of purchasing the lines on payment of an annuity for the unexpired portion of the ninety-nine years' concession, calculated on the average profits of the lines during the seven years previous to the act of taking.

The French method of operating the railroads is as far removed from the English or American as is the system under which they were constructed. The supervision of the government is ubiquitous. Every tariff, every time-table, has to be submitted for approval, and there are public agents at every principal station. The accounts of the companies are subjected to an annual examination, and the most rigid police regulations are enforced. If questions arise between companies, they are settled not by might asserting itself through competition, but by a board of arbitration, with an ultimate appeal in matters of graver importance to the Central Railroad Commission.

Thus it is that, in theory, the railroad system of France is purely and essentially French. The government initiated it, supervises it, has a large ultimate pecuniary interest in it. At the expiration of sixty years more it may yet be made to pay off the national debt. At present, however, it is accumulating it. The guaranteed interest is a constant burden on the revenue. And it is in this connection that the French railroad problem asserts itself. The essence of the system lies in regulation, as a substitute for competition. One railroad war, such as annually vexes America, would make the guaranty of the government assume proportions calculated to appal the most daring minister of finance. One can imagine the fury of American railroad struggles if the payment of interest was guaranteed from the public treasury! Competition, therefore, cannot be tolerated among the railroads of France. The French public, nevertheless, like the English and the

American, is constantly demanding more railroads. It asks for them, too, not because they are profitable in themselves, but because of the incidental advantages to be derived from them. The great established companies naturally say that there must be some limit to construction. They can ruin neither themselves nor the government by building railroads intended merely to improve the value of adjacent property. To this those demanding the additional roads simply reply that if the great companies will not supply them, they desire the privilege of supplying themselves.

Yielding to this plausible argument, and to a feeling of political necessity, a law of the empire, known as the railroad law of the 12th of July, 1865, undertook to create a third *réseau* called the *réseau vicinal*. It was a French approach to the American idea of a general railroad law. The departments and communes were empowered either to construct certain local railroads themselves, or to grant charters for their construction by others. It was erroneously supposed that these roads would be insignificant affairs, and act as mere feeders to the great companies. The French do not move rapidly in enterprises of this description, but still they move. The door was now open; competition soon entered through it. At first few local concessions were made, and those in good faith. Then the projects began to flow in, and they rapidly assumed a new phase. The contractor, the speculator, and the black-mailer made their appearance in rapid succession. Railroads were built to be sold.

A new character now suddenly appeared upon the stage in the person of a daring Belgian operator, or railroad king, one Phillipart by name. This man had seen his opportunity some years before in Belgium, and by constructing and connecting a number of cheap railroads in that country parallel with the state roads he had succeeded in so embarrassing the operations of the latter, that the government had in 1870 leased his properties on terms very advantageous to him. It in fact bought him out. Transferring himself to Paris he there attempted a similar operation on the French roads under cover of the *réseau vicinal*,—connecting the disconnected lines into competing systems. He wielded an enormous capital and operated on a large scale. For a time he occasioned the government and the old established lines much perplexity. Obtaining control of several banks and taking advantage of the mania for railroad construction, he developed his plans with great rapidity. In 1870, the local lines constructed under the law of 1865 aggregated but 180 miles.. During the war with Germany the amount did not, of course, increase. Under the stimulus of the Phillipart mania, however, it rapidly assumed new proportions, and in 1875 there were 930 miles of completed local roads, while 1730 more miles were in process of actual construction, and 756 miles authorized. These roads involved an estimated outlay of $130,-

000,000. The posture of affairs was highly alarming, and the government was being rapidly forced into a very difficult position. The French nation had a large pecuniary interest in the existing railroad system. It was in fact under the terms of the concessions a vast sinking fund for the future extinction of the national debt. Its value was now seriously menaced by that uncontrolled competition which it had been the whole effort of the French railroad policy to provide against. Yet the hands of the government were tied. It did not dare to run counter to the mania and repeal the law of 1865, for that would have been tantamount to forbidding all future railroad construction. It seemed, indeed, as if the Phillipart scheme must again prove a success, and that the established lines would have to submit to being victimized. The mania, however, did not suffice to overcome the obstacles in the great operator's way. The odds against Phillipart were too heavy. He was broken down in his stock speculations by a general combination against him, and failed for an enormous amount early in 1876. With his failure the mania collapsed. The question, however, only assumed a new shape. The local lines were worthless in themselves, but those holding stock in them were not without influence, and they now turned to the government. The question of the purchase of all the railroads by the state was agitated; and it was claimed the government should at least complete the unfinished local roads and do something to relieve their owners. The plan of purchase by the state encountered, however, much resistance from the existing companies, and the discussion would probably soon have died away, had it not recently been revived through the publication of two reports prepared by the present Minister of Public Works. Here the matter now rests and the course which events will take is not apparent. Meanwhile, the whole drift of discussion tends, away from the private ownership of French railroads and reliance on competition among them, towards a closer connection between the railroads and the government. So far as uncontrolled competition is concerned, the result of the Phillipart struggle has, indeed, been decisive. It will hardly again be seriously attempted. On the other hand the vested interests in the established companies are so powerful that it seems improbable they will be disturbed. The relations between the community in France and its railroad system are moreover reasonably satisfactory, and no strong disposition to force a change is apparent. Though not especially enterprising, the companies are as a rule solvent, impartial and reliable. Indeed those managing them look with simple astonishment on the wild fluctuations in the railroad tariffs incident to the American method of operation, and they do not hesitate to say that if any similar outrages were perpetrated on the French people and business public by them, the question of the state ownership of railroads would immediately assume a new shape. Such proceedings would not be tolerated.

If there is, indeed, an inherent and irresistible tendency in the railroad systems of all countries to assume closer relations with governmental systems; if, as so many are inclined to believe, transportation is such an important and complex element in modern life that it must ultimately find its place among the functions of the state, then it is safe to say that in no other country does the railroad problem present so interesting a phase of present development as in Germany. The inclination of the German mind, especially the North German mind, is bureaucratic. It takes naturally and kindly to this method of development. It seems the natural mode in which the political genius of the people works. With us, in America, it is just the opposite. The commission is our bureau. We are continually driven to a recourse to it, but we always accept the necessity with reluctance, and the machine withal does not work well. Where it is not corrupt, it is apt to be clumsy. We get from it no such results as are obtained by the Germans. The reason, if we choose to seek it, is obvious enough. The bureau is a natural outgrowth of the German polity; it is the regular and appropriate form in which that polity effects its work. With us it is a necessity, but none the less an excrescence. Our political system has come in contact, through the complex development of civilization, with a class of problems in presence of which it has broken down; such questions as those of police, sanitary regulations, education, internal improvements, transportation. At first we always try to deal with these through the machinery of parliamentary government, a sort of sublimated town meeting. The legislative committee is the embryotic American bureau; as such it serves its purpose for a time, doing its work in an uncouth, lumbering sort of way, and then, its insufficiency becoming manifest, it makes way for the commission. The American commission is, however, by no means the Prussian bureau. It is at best a very poor substitute for it; a thing suddenly improvised in place of one gradually developed.

When a community is brought face to face with such a problem as the gradual political development, it might almost be said the political evolution, of its railroad system, this distinction becomes important. In the one case the question is approached by a patient, trained professional; in the other by an eager, ever-confident amateur. If, therefore, the problem of reëstablishing the state in new and more effective relations with the agencies of transportation is to be solved in our time, it is pretty safe to predict that the solution will be reached in Germany long before it is in America. Not only do they approach it there in a more practical and scientific spirit, but the ground is better prepared. The material is more ready to the hand. For, almost necessarily, the German railroad system reflects the condition of the German political system. It is a curious complication, very difficult to understand:—a mass of raw material, out of which order is to be deduced. Particularism rules supreme; each petty sovereignty has a policy of its own. Yet certain fundamental principles have asserted themselves everywhere. The system, for instance, was originally established on the principle of concessions to private companies, usually for from thirty to fifty years, and the idea of competition found no place in it. On the contrary, the building of competing lines was expressly forbidden. As the several lines extended themselves, this restriction so impeded their development that in Prussia a few years ago it was repealed. The results which have just been described in France then ensued. A mania for railroad construction and expansion developed itself. Dr. Strausburg burst upon an astonished world. The usual result followed. A panic and collapse took place, and railroad property depreciated in value as much in Prussia as recently it has in America.

But throughout Germany the relations between the state and the railroads have always been very close. Those building the roads under concessions have received liberal aid from government, sometimes in the form of a subsidy, at other times through a guaranty of interest or dividends; while in yet other cases the state itself has been a large stockholder. The tendency towards a closer connection between the government and the railroads has constantly been apparent, and is more pronounced now than ever before. Prussia, always a large, if not the largest, owner and manager of railroads in North Germany, has lately purchased new lines; while the government of Bavaria has at last acquired all the railroads within the limits of that country, and is indeed thus the first considerable government in the world to both own and work its entire system. Whether actually owning and operating the railroads or not, however, the hand of the German governments has ever been present in their affairs, regulating everything, from the rates on merchandise to the safeguards against accident. Starting from the fundamental German principle that it is not only the right but the duty of the state to interfere in every matter of public interest, it assumed the power as a matter of course, until in practice the will of the minister was able to make itself felt in every direction.

Owing to the lack of cohesion among the political organizations of the German-speaking race, the necessities of their position long ago caused the railroads of central Europe to form a union among themselves. In this there were included, in 1873, nearly one hundred managements, operating 26,000 miles of track, the governments being represented in the same way as private managements. This union settled questions of fares and freights, and made all necessary traffic arrangements. Through it combination was made to take the place of competition, and in case of controversy the roads had recourse to arbitration, directly under the eye of the government and of the public, instead of to wars of rates. Before the battle of Sadowa brought the North German empire into existence, this union was, under the conditions there existing, a necessity. It then became firmly established, and is now recognized as a most useful part of the railroad organization. It introduces into the system uniformity and stability, causing a direct contact with the government.

Holding in ownership and themselves operating the whole or a large portion of the railroads within their limits, as so many of the German governments did, it was inevitable that the formation of the German empire must here also work a new departure. The control of the lines of communication was very essential to the stability of the new imperial system. Provision in this respect was accordingly made in the constitution of the empire, and the policy to be pursued under it has since been not the least perplexing of the many perplexing questions which have engaged the attention of Prince Bismarck. At first he seemed to incline towards a scheme of general state ownership. All the railroads were by degrees to be purchased and absorbed by the imperial government. As a step in this direction a commission was appointed which in 1876 made an elaborate report. This was submitted to the Prussian parliament, accompanied by a government project for the transfer of the state roads of that country to the empire. After a vigorous debate, in which Bismarck himself took an active part, the measure was passed, though not without considerable opposition. Nothing, however, has been done under it; on the contrary the movement seemed to stop here, and it is generally supposed will not, at any rate for the present, be pressed further. It is further stated that Bismarck after full consideration has come to the conclusion that the correct solution of the problem does not lie in the direction of state ownership. He has complained bitterly in the Prussian parliament of repeated conflicts between the imperial government and the managements of the railroads, and not long since proposed to create a special ministry to take charge of the subject so far as Prussia was concerned. This measure was defeated, and led to a change in the cabinet. Meanwhile Baron von Weber, who is the highest German authority on railroad subjects, has recently entered the Prussian service, and is thus in position to affect the course of affairs. He is committed against the project of imperial ownership, and inclines to the adoption of a definite policy very similar to that which has been arrived at in Great Britain. He would leave the system to develop in its own way, and to assume such shape as circumstances may dictate. Those owning the individual railroads, whether states or companies, should be left to manage them, provided they did so under close governmental supervision. In Germany it should be understood this means much more than it does in Great Britain, or would in America. Meanwhile the German policy seems to incline towards at least trying the experiment of government control, before having recourse to government ownership. The whole question there, however, is complicated by political and military considerations which do not affect it either in Great Britain or in America.

The English, the Belgian, the French, and the German are the four great railroad systems. With many points in common, each has peculiar features deserving of careful study. In their political relations they are divided into two groups by a broad line of demarkation. On the one side of that line are the systems of the English-speaking race, based upon private enterprise and left for their regulation to the principles of *laissez faire*, the laws of competition, and of supply and demand. On the other side of the line are the systems of continental Europe, in the creation of which the state assumed the initiative, and over which it exercises constant and watchful supervision. In applying results drawn from the experience of one country to problems which present themselves in another, the difference of social and political habit and education should ever be borne in mind. Because in the countries of continental Europe the state can and does hold close relations, amounting even to ownership, with the railroads, it does not follow that the same course could be successfully pursued in England or in America. The former nations are by political habit administrative, the latter are parliamentary; in other words, France and Germany are essentially executive in their governmental systems, while England and America are legislative. Now the executive may design, construct, or operate a railroad; the legislative never can. A country, therefore, with a weak or unstable executive, or a crude and imperfect civil service, should accept with caution results achieved under a government of bureaus. Nevertheless, though conclusions cannot be adopted in the gross, there may be in them much good food for reflection.

The railroad system of the United States, with all its excellences and all its defects, is thoroughly characteristic of the American people. It grew up untrammelled by any theory as to how it ought to grow; and developed with mushroom rapidity, without reference to government or political systems. In this country alone were the principles of free trade unreservedly and fearlessly applied to it. The result has certainly been wonderful, if not in all respects satisfactory. Why it has not been wholly satisfactory remains to be explained.

Looked upon as a whole, the American railroad system may now be said to have passed, wholly or in part, through three distinct phases of growth the limits of which are merged in each other, though their order of succession is sufficiently clear. First was the period of construction, beginning with the year 1830 and closing with the completion of the Pacific railroad in 1869;—merged with this period and following upon it, was that of active competition, which reached its fullest development in 1876;—this naturally was followed by the period of combination, which first assumed a large and definite shape in 1873, and has since that gradually but surely been working itself out into something both definite and practical. To one now looking back and reviewing the whole course of events, cause and effect become apparent. Things could not have taken any course other than that which they did take,—the logic has been inexorable. The whole theory under which the railroad system was left to develop itself was founded on a theoretical error; and it was none the less an error because, even if it had been recognized as such, it could not have been remedied. That error lay in the supposition, then universally accepted as an axiom, that in all matters of trade, competition, if allowed perfectly free play, could be relied upon to protect the community from abuses. The efficacy of railroad competition,—expressing itself in the form of general laws authorizing the freest possible railroad construction everywhere and by any one,—at an early day became almost a cardinal principle of American faith.

The people of the country in their political capacity had faith in it. Indeed, not to have had faith in it at that time would have seemed almost to imply a doubt of the very principles upon which the government was established. The whole political experiment in America was based upon the theory that the government should have the least possible connection with all industrial undertakings, that these undertakings had been regulated in other countries far too much, and that now, in the New World, it was to be proved that they would regulate themselves best when most left alone. The exceptions to this rule had yet to develop themselves. Forty years ago they had not begun, or had hardly begun, to develop themselves at all. If the people, and through the people the government, had faith in competition, the private individuals who constructed the railroads seemed to have no fear of it. They built roads everywhere, apparently in perfect confidence that the country would so develop as to support all the roads that could be built. Consequently railroads sprang up as if by magic, and after they were constructed, as it was impossible to remove them from places where they were not wanted to places where they were wanted, they lived upon the land where they could, and, when the business of the land would not support them, they fought and ruined each other.

The country was of immense extent, and its development under the stimulus of the new power was unprecedentedly rapid. At first, and during the lives of more than one generation, it really seemed as if the community had not relied upon this fundamental law of competition without cause. Nevertheless, there never was a time, since the first railroad was built, when he who sought to look for them could not find in almost any direction significant indications of the violation of a natural law. Local inequalities always existed, and the whole system was built up upon the principle of developing competing points at the expense of all others. There were certain localities in the country known as railroad centres; and these railroad centres were stimulated into an undue growth from the fact that competition was limited to them. The principles of free trade did not have full play; they were confined to favored localities. Hence resulted two things: in the first place the community suffered; then the railroads. Under the hard stress of local and through competition the most glaring inequalities were developed. The work of the railroad centres was done at a nominal profit, while the corporations recompensed themselves by extorting from other points where competition did not have to be met, the highest profit which business could be made to pay. It thus gradually became apparent, although men were very slow to take in the fact, that immense and invaluable as were the results in many respects secured through unlimited railroad competition, yet so far as the essential matter of securing to all reasonable and equal rates of payment for similar services performed was involved, it did not produce the effect confidently expected of it. On the contrary, it led directly to systematic discriminations and wild fluctuations, and the more active the competition was, the more oppressive the discriminations became and the less possible was it to estimate the fluctuations. In other words, while the result of other and ordinary competition was to reduce and equalize prices, that of railroad competition was to produce local inequalities and to arbitrarily raise and depress prices. The teachings of political economy were at fault. The variation was so great that it was evident some important factor in the problem had been overlooked.

Though in the case of the railroad system the disturbance produced by this false application of a correct principle was far more sensibly felt in America than in any other country, yet the reason of the difficulty was thought out elsewhere. Much has been heard within the last few years of a newly developed school of political and economic thinkers which is making itself felt in Germany, and the tendency of which is supposed to be reactionary against English free trade and *laissez faire*. These German thinkers have been laid hold of in this country by the protectionists, and claimed by them as allies. In truth they are nothing of the sort. They are free traders themselves, but they declare that the principles of free trade also are not of unlimited application; that, on the contrary, experience, and especially the experience of the last few years, has definitely shown that, in the complex development of modern life functions are more and more developed which, in their operation, are not subject to the laws of competition or the principles of free trade, and which indeed are reduced to utter confusion within and without if abandoned to the working of those laws. The more thorough ascertainment of these limitations on principles generally correct is one of the important studies of the day. Thirty years ago they were not understood at all; they are now understood only in part. John Stuart Mill had a clear though limited perception of them; and how limited his perception was will be realized from the fact that of the twelve hundred pages of his work on political economy he devotes just four pages to this subject. Yet to-day these limitations are asserting themselves in a way which cannot be ignored.

The traditions of political economy, therefore, to the contrary notwithstanding, there are functions of modern life, the number of which is also continually increasing, which necessarily partake in their essence of the character of monopolies. The supplying railroad and telegraph facilities presents examples of these on the large scale, as the supplying cities and large towns with water and gas presents examples of them on a smaller scale. All of these, and others which could be named, partake of the character of monopolies as a fundamental condition of their development. Now it is found that, wherever this characteristic exists, the effect of competition is not to regulate cost or equalize production, but under a greater or less degree of friction to bring about combination and a closer monopoly. The law is invariable. It knows no exceptions. The process through which it works itself out may be long, but it is sure. When the number of those performing any industrial work in the system of modern life is necessarily limited to a few, the more powerful of those few will inevitably absorb into themselves the less powerful.

The difficulty of the process is a mere question of degree; its duration is a mere question of time. In America a great many agents, though by no means an unlimited number, are employed in the work of railroad transportation, hence the monopoly is looser and the struggle between the monopolists is fiercer than it is in many other countries; hence, also, the process of bringing about a thorough combination is rendered more difficult and requires more time. None the less it goes on.

Where the extent of country to be occupied was so vast and the necessity for some means of transportation so great, it naturally took a number of years for a theoretical error at the bottom of a system to work its way to the top. For a long time all went apparently well. The people of the country saw only the manifold advantages which flowed from a railroad construction which was stimulated by every inducement which could be held out to avarice. Thousands of miles were built each year,—the interior was opened to the seaboard with an energy which outdid expectation,—new appliances, whether of speed, of safety or of economy were introduced as fast as ingenuity could invent them,—rates of fare and of freight between distant points became lower and lower, until what seemed reasonable yesterday was looked upon as exorbitant to-day,—and altogether the development was so surprising that it could not but excite sensations of wonder and gratitude which for the moment alone found expression. This state of things could not be permanent. The mania for railroad construction which began in 1866, and culminated in the crash of 1873, brought matters to a crisis. As lines multiplied, the competition increased. The railroads had been built much too rapidly and the business of the country could not support them. Those immediately in charge were under a constant and severe pressure to earn money; and they earned it wherever and however they could. They stopped at nothing. Between those years it is safe to say that the idea of any duty which a railroad corporation owed to the public was wholly lost sight of. In the eyes of those managing them the railroads were mere private money-making enterprises. They acted accordingly. If they were forced to compete, they competed savagely and without regard to consequences;—where they were free from competition, they exacted the uttermost farthing. There naturally ensued a system of sudden fluctuations and inequitable local discriminations which has scarcely ever been equalled and which was well-nigh intolerable. At one point several roads would converge, and the business or travel to and from that point would be furiously fought over until rates became almost literally nominal; meanwhile those engaged in business or living at other points but a few miles away would be charged every penny that they could be made to pay without being driven off the railroad and back to the highway. Where goods starting from the same point were to be delivered at different stations on the line of the same road, those forwarding them discovered to their cost that the tariff resembled nothing so much as an undulating line,—for a distance of twenty miles, more would have to be paid than for a distance of forty miles. Those living between competing points were rigidly excluded from the benefits of competition. To such an outrageous extent was this carried, that it became the common practice where an entire car-load of merchandise, destined to some way station on the line of a railroad, was paid through to a competing point far beyond on that line, to make a large extra charge for *not* hauling it to that point, but dropping it at its ultimate destination in the first place. It was exactly as if a traveller was to buy a through ticket from Albany to Buffalo, and the railroad company were to insist not only on taking up his ticket but on charging him a dollar extra if he left the train at Syracuse. Remonstrances against this absurd anomaly were treated by the officials of the railroad companies as if they were too unreasonable for a patient hearing. This was much more the case in the West than at the East, but even in Massachusetts when the state railroad commissioners on one occasion urged upon a corporation the injustice of charging some twenty dollars extra on each car-load of wheat which had been paid for on a through bill of lading to the further end of its line if they left it at its point of destination one hundred miles short of that end, thus making a charge of twenty dollars for *not* hauling the loaded car one hundred miles,—in this case their representations were met with the counter proposal that, if the consignee preferred, the company would haul his goods by his door to the point to which they were billed, and then back, charging both ways. He might pay the through rate forward and the local rate back, or submit to the extra charge, just as he chose. Besides all this, however, competition led to favoritism of the grossest character,—men or business firms whose shipments by rail were large could command their own terms, as compared with those whose shipments were small. The most irritating as well as wrongful inequalities were thus made common all over the land. Every local settlement and every secluded farmer saw other settlements and other farmers more fortunately placed, whose consequent prosperity seemed to make their own ruin a question of time. Place to place, or man to man, they might compete; but where the weight of the railroad was flung into one scale, it was strange indeed if the other did not kick the beam.

Of course, even under the most favorable circumstances, it was wholly unlikely that such a condition of affairs should long continue. The fact that these abuses were the simple and inevitable outcome of a public policy in regard to railroads which had from the beginning been jealously adhered to was of no consequence. People felt and did not reason. Competition made the price of flour and cloth and shoes equal and reasonable: why should it make fares and freights unequal and unreasonable? Few indeed were they who could be made to see that the true cause of complaint was with an economical theory misapplied, not with those who with only too much energy had carried out the misapplied theory to its final logical conclusions. Yet a cause of complaint did exist, and to a degree which made a popular explosion inevitable. In the case of the railroad corporations, moreover, the prejudice was aggravated by well authenticated rumors of the gross financial scandals which disgraced their management. The system was, indeed, fairly honeycombed with jobbery and corruption. They began high up in the wretched machinery of the construction company, with all its thimble-rig contrivances to effect the unseen transfer of assets from the treasury of the corporation to the pockets of its directors. Thence they spread downward through the whole system of supplies and contracts and rolling-stock companies, until it might not unfairly be said that everything had its price.

The natural results followed. In 1870 a popular agitation broke out which for the time being threatened to sweep down not only all legal barriers but every consideration of self-interest; and, at the same time, the corporations driven to the verge of bankruptcy, if not fairly over it, by the joint effects of corruption and competition, turned their thoughts on the single chance of escape which combination among themselves held out. In 1873 the Grangers were electing Judges in Illinois; and in 1874 the railroad magnates were discussing the details of a grand combination at Saratoga.

Of the Granger episode little now needs to be said. That it did not originate without cause has already been pointed out. It is quite safe to go further and to say that the movement was a necessary one, and through its results has made a solution of the railroad problem possible in this country. At the time that movement took shape the railroad corporations were in fact rapidly assuming a position which could not be tolerated. Corporations, owning and operating the highways of commerce, they claimed for themselves a species of immunity from the control of the law-making power. When laws were passed with a view to their regulation, they received them in a way which was at once arrogant and singularly injudicious. The officers entrusted with the execution of those laws they contemptuously ignored. Sheltering themselves behind the Dartmouth College decision, they practically undertook to set even public opinion at defiance. Indeed there can be no doubt that those representing these corporations had at this juncture not only become fully educated up to the idea that the gross inequalities and ruinous discriminations to which in their business they were accustomed were necessary incidents to it, which afforded no just ground of complaint to any one; but they also thought that any attempt to rectify them through legislation was a gross outrage on the elementary principles both of common sense and of constitutional law. In other words, they had thoroughly got it into their heads that they as common carriers were in no way bound to afford equal facilities to all, and, indeed, that it was in the last degree absurd and unreasonable to expect them to do so. The Granger method was probably as good a method of approaching men in this frame of mind as could have been devised. They were not open to reason, from the simple fact that their ideas of what in their position was right or wrong, reasonable or unreasonable, were wholly perverted. They were part of a system founded on error; and that error they had all their lives been accustomed to look upon as truth. The Granger violence was, therefore, needful to clear the ground. This it did; and it did it in a way far from creditable to those who called themselves Grangers.

Indeed, the extravagant utterances of that time would even now seem incredible were they not matter of record. For instance, the following is one of a long series of resolutions adopted at a general convention of the Granges held at Springfield, Ill., on the 2d of April, 1873:—

Second. The railways of the world, except in those countries where they have been held under the strict regulation and supervision of the government, have proved themselves of as arbitrary extortion and opposed to free institutions and free commerce between the states as the feudal barons of the Middle Ages.

This comparison between the modern railroad corporations and the feudal barons, in spite of its grotesque absurdity, was very popular among the Granger rhetoricians. It made its appearance with great regularity in nearly all their more labored and ornate productions. In June, 1873, for example, numerous county gatherings put forth a declaration of farmers' grievances and principles, in which occurred this passage:

"The history of the present railway monopoly is a history of repeated injuries and oppressions, all having in direct object the establishment of an absolute tyranny over the people of these States unequalled in any monarchy of the Old World, and having its only parallel in the history of the mediæval ages, when the strong hand was the only law, and the highways of commerce were taxed by the feudal barons, who, from their strongholds, surrounded by their armies of vassals, could lay such tribute upon the traveller as their own wills alone should dictate."

As usual, these wild utterances in due time resulted in the enactment of yet wilder laws. Laws were demanded which should regulate the profits, the methods of operation, and the political relations of the railroads; the corporations were to be made to realize, as the phrase went, that "the created was not greater than the creator;" that the railroads were the servants of the people and not their masters. Here then ought to have been met a complete and logical abandonment of the whole theory of regulation by natural law, under which the railroad system had been organized and had grown up. If that theory was worth anything at all, the remedy for the ills under which the community was suffering would at once come into play. The railroads were not monopolies. There was nothing to prevent the organization of new companies to construct parallel and competing lines of road. Here was the remedy through competition :—and the mere statement of it revealed its utter absurdity. Nevertheless the idea that from the very necessity of the case uncontrolled railroad competition led directly to and was inseparable from railroad discriminations and local inequalities obtained no lodgment.

The fact that the railroad companies did not compete with each other regularly and equally and moderately at all times and at all places was patent. The reason why they did not do so was not at once apparent, and the result itself, therefore, by a sort of general consent, was set down as one more manifestation of that innate perversity common to all monopolists. Meanwhile not the slightest degree of distrust was felt of the competitive principle under the conditions in which it was now sought to be applied. That, throughout the discussion, was accepted as axiomatic—a nostrum at once universal and infallible.

The abandonment of competition between railroads consequently found no place in the philosophy of Granger legislation. It was, on the contrary, tenaciously clung to, and new laws were passed to render more illegal than ever any combination between competing lines. The remedy, from the Granger point of view, was obvious. As the trouble was due to human perversity, and not to any defect in principle, nothing was needed to make things right but more of the same remedy, liberally supplemented by penal legislation. If the Sangrado treatment did not work a cure, the blood-letting must go on, but the alguazil should replace the warm water.

Meanwhile the Granger legislation, crude as it was and utterly as it lacked insight, did produce results. That it did so was due wholly to the fact that the states which enacted the Granger laws went further, and incorporated into them a special executory force. To a certain extent, therefore, the state governments assumed the management of the railroads. In so far as they did this, the Granger legislation was logical and consequently effective. Government regulation is a practical substitute for competition. Apart from this, it was in no respect a success. If experience has proved anything conclusively, it has certainly proved that mere abstract laws aimed at the inequalities which arise out of railroad competition are of no avail. Whether placed on the statute books as laws generally applicable, or incorporated at length into special charters, the result has been the same. The precedents are innumerable, and the Granger experiments did but add to their number. The ingenuity of lawyers, working on the intricacies of a most complicated system, has never failed to make a broad path through the meshes of merely declaratory statutes. The Granger legislatures, though with great reluctance, recognized this fact. Boards of commissioners were accordingly provided, and to them was entrusted a general supervision over the railroads and the duty of making the new legislation effective. The organization and experience of these commissions is, from the governmental point of view, the most important and instructive phase in the development of the railroad problem during the last few years.

It has already been pointed out that the inclination of the American mind is not bureaucratic. Recourse is had in this country to commissions, as our bureaus are called, with great reluctance. Experience, it must also be admitted, fully justifies this feeling of distrust; as a rule they do not work well. Not only do they develop in too many cases a singular aptitude for all jobbery, but, even when honestly composed, they rarely accomplish much. Once created, also, they can never be gotten rid of. They ever after remain part of the machinery of government, drawing salaries and apparently making work for themselves to do. The reason is obvious. In America there are not many specialists, nor have the American people any great degree of faith in them. The principle that all men are created equal before the law has been stripped of its limitations, until in the popular mind it has become a sort of cardinal article of political faith that all men are equal for all purposes. Accordingly, in making up commissions to deal with the most complicated issues arising out of our modern social and industrial organization, those in authority are very apt to conclude that one man can do the work about as well as another. The result is what might naturally be expected; and the system is made responsible for it. All this received pointed illustration in the case of the Granger commissions. In the first place the country did not contain any trained body of men competent to do the work. They had got to be found and then educated. In the next place the work was one of great difficulty and extreme delicacy. The commissioners were to represent the government in a momentous struggle with the most compact and formidable interest in the country. They were to be pitted against the ablest men the community could supply, thoroughly acquainted with their business and with unlimited resources at their disposal. Finally the test of success was to be that, under these circumstances and in the face of these difficulties, the commissioners should develop the crude original laws placed in their hands into a rational and effective system.

It was from the beginning, therefore, obvious that no high standard of success could reasonably be hoped for from the Granger commissions. They were far too heavily handicapped. In the first place the executives of the states in selecting their members not infrequently seemed to regard any antecedent familiarity with the railroad system as a total disqualification. So afraid were they of a bias, that they sought out men whose minds were a blank. Farmers, land-surveyors, men of business and politicians were selected. There were, of course, exceptions to this remark, and some very competent men were appointed who did excellent work so long as they remained in office. But a long continuance in office was again looked upon as undesirable, and these men were either speedily removed to make way for incompetents, or they voluntarily passed into the employ of the railroad corporations before they had fairly mastered the situation.* Above and beyond all this, however, these commissions began their work in a false position, and they never extricated themselves from it. They were not judicial tribunals. They ever reflected the angry complexion of the movement out of which they had originat-

* There could not be a better illustration of the shifting character of these boards under the system in use in the states of the West than has been furnished in the case of that of Illinois. It consists of three members. The original appointments were made July 1st, 1871. These commissioners went out of office and an entirely new set were appointed on March 13th, 1873. The chairman of the new board died in the succeeding November, and a successor was appointed. These commissioners held office until February 21st, 1877, when they all retired, and were succeeded by the present board. Three complete changes in less than six years, with one additional vacancy occasioned by death. Under these circumstances a remark in its last report that "the Commission, ever since the time of its organization, has labored under embarrassments which have deprived it of the ability to be as useful to the people" as it might have been, seems in no way unreasonable. But who is responsible? Certainly not the commissioners, who by no possible exertion of their own during their brief tenure of office could have qualified themselves to perform its duties. The railroad corporations manage things differently.

2*

ed. They were where they were, not to study a difficult problem and to guide their steps by the light of investigation. Nothing of this sort was, as a rule, expected of them. On the contrary they were there to prosecute. The test of their performance of duty was to be sought in the degree of hostility they manifested to the railroad corporations. In a word they represented force.

That under these circumstances they succeeded at all is the true cause of astonishment; not that they succeeded but partially. That they did succeed was due solely to the incorrigible folly and passionate love of fighting which seems inherent in the trained American railroad official. Placed as they were, entirely unfamiliar with the difficult questions they were compelled to confront, lacking confidence in themselves and very much afraid of their opponents, had those opponents seen fit to be even moderately civil and deferential to them, the position of the commissioners would have been rendered extremely difficult. Had the representatives of the railroad corporations, with their vast resources and intimate acquaintance with the subject, been wise enough to take the initiative and meet the commissioners half way, it would have been strange indeed if they had not succeeded in impressing upon them a sense of the difficulty of their task, and so materially affected their action. Instead of this they simply ignored them. For instance, when the newly organized California board requested one corporation to forward the passes which would enable the commissioners to go over its road, the passes came accompanied by a denial of the "validity of the law," and a statement that they were sent "not in obedience to said Act, but merely as an act of courtesy to the members of the Board and their Secretary." Even in Massachusetts, the mere suggestion of the commissioners in 1871 to the railroad corporations that they should carefully revise their tariffs, was met by one General Manager with the astounding reply that "he had not supposed, and did not now suppose that the Commission intends to seriously attempt advising the trained and experienced managers of roads in this Commonwealth upon the details of their duty." In the West, during the years 1872–3, if a railroad official was asked what course the companies proposed to pursue in regard to the new legislation, the usual answer was that they did not propose to pay any attention whatever to it. Imagine the English corporations thus coolly setting Parliament at defiance! Naturally this impolitic course not only incensed the commissioners, but, what was of far more consequence, it strengthened their hands. The popular feeling, strong enough before, was intensified. The agitation was thus kept alive until the decision of the courts of last resort was obtained, which fortunately placed the railroads completely at the mercy of the legislatures. Nothing short of this would apparently have sufficed to force them out of their attitude of stupid, fighting defiance. This re sult, however, once arrived at, they immediately recovered their senses, and with them their strength. They became at once compliant and—formidable. Their knowledge, their skill, their money and their influence began to tell. Driven by brute force out of the utterly untenable position in which they had sought to entrench themselves, the moment they reconciled themselves to the use of the weapons of protection customary in civilized communities matters began sensibly to improve. Not that the problem was touched, for discrimination and inequality, competitive business and local combinations, still remained inherent in the system. An obligation to the public was, however, recognized. It was no longer claimed that railroads were mere private business enterprises, and the abuses incident to their competition among themselves were at least softened down by the absence of that old arbitrary spirit which had so aggravated hardships. The laws were sufficiently complied with to remove the more flagrant causes of complaint, and the practical results thus secured through the Granger agitation were far more considerable than has been generally supposed.

Fortunately, while in the more western states of the Union years were being wasted in a mere preliminary struggle, the question in another part of the country had from the beginning taken a different shape, and one far more promising of results. Owing to other conditions of railroad ownership and a more composed state of the public mind, the East afforded a better field for profitable discussion than the West. Various state railroad commissions already existed in that section, but in 1869 one was organized in Massachusetts on a somewhat novel principle, and a principle in curious contrast with that which has just been described as subsequently adopted in the West. In the West the fundamental idea behind every railroad act was force;—the commission represented the constable. In the Massachusetts act the fundamental idea was publicity;—the commission represented public opinion. The law creating the board and defining its field of action was clumsily drawn, and throughout it there was apparent a spirit of distrust in its purpose. In theory an experiment, in reality it was a makeshift. The powers conferred on the commissioners hardly deserved the name; and such as they were, they were carefully hedged about with limitations against their abuse. Accordingly when the commissioners entered upon their duties they were at first inclined to think that they could hardly save themselves from falling into contempt from mere lack of ability to compel respect for their decisions. In fact, however, the law could not have been improved. Had it not been a flagrant legislative guess, it would have been an inspiration. The only appeal provided was to publicity. The board of commissioners was set up as a sort of lens by means of which the otherwise scattered rays of public opinion could be concentrated to a focus and brought to bear upon a given point.

The commissioners had to listen, and they might investigate and report;—they could do little more. Accordingly they were compelled to study their subject, and with each question which came before them they had to stand or fall on the reasons they presented for their conclusions. They could not take refuge in silence. Whenever they attempted to do so they speedily found themselves in trouble. They had, as each case came up, to argue the side of the corporations or of the public, as the case might be; but always to argue it openly, and in a way which showed that they understood the subject and were at least honest in their convictions. Placed from the beginning in this position, the board was singularly fortunate in the permanence with which its members were continued in office. But two individual changes were made in it during nine years, and it has undergone no change during the last six. Accordingly it had a chance to outlive its inexperience and profit by its own blunders, which naturally were at first neither trifling nor infrequent. The result was necessarily as different from that reached at the West, as were the conditions under which it was reached. The board, in the first place, became of necessity a judicial in place of a prosecuting tribunal. It naturally had often to render decisions upon matters of complaint which came under its cognizance in favor of the railroad corporations;—whether it decided in their favor or against them, however, its decisions carried no weight other than that derived from the reasons given for them. The commissioners were consequently under the necessity of cultivating friendly relations with the railroad officials, and had to inspire them, if they could, with a confidence in their knowledge and fairness. Without that they could not hope to sustain themselves. On the other hand, their failure was imminent unless they so bore themselves as to satisfy the public that they were absolutely independent of corporate influence, and could always be relied upon to fearlessly investigate and impartially decide.

Undesignedly the Massachusetts legislators had rested their law on the one great social feature which distinguishes modern civilization from any other of which we have a record,—the eventual supremacy of an enlightened public opinion. The line of policy thus happily initiated was carefully pursued. New and wider powers were, year by year, conferred upon the board, but always in the same direction,—powers to investigate and report. The commissioners meanwhile were not slow to realize the advantage of their position, and have repeatedly put themselves on record as desiring no more arbitrary powers,—as feeling themselves indeed stronger without them. In 1876, this policy reached its final result, as the legislature then placed the entire system of accounts kept by the corporations under the direct supervision of the board. Its power in this respect was unlimited. Not only was it authorized to prescribe a uniform system upon which those accounts should be kept, but they were also to be kept under the immediate and constant supervision of its officers, and on proper application the books were to be publicly investigated. In view of the notorious scandals which have made "railroad financiering" a by-word for whatever is financially loose, corrupt and dishonest, the scope and significance of this measure does not need to be dwelt upon. It went to the root of the matter. It opened to light all the dark places. In France only, it is believed, had a similar power been asserted; but there, its exercise was based on the large pecuniary interest the government had in the railroad properties. It was a partner, and as such concerned in all their transactions. In Massachusetts a different ground was taken. The indisputable fact was recognized that those corporations are so large and so far removed from the owners of their securities, and the community is so deeply concerned in their doings and condition, that the law-making power both has a right and is in duty bound to insist on that publicity as respects their affairs without which abuses cannot be guarded against. No where has the soundness of this doctrine received such copious illustration as in America during the last few years. Singularly enough also, this act was passed not only without opposition from the railroad companies as a body, but with the active assent of many of them. When it took effect the corporations were summoned together by the commissioners and invited to assist, through a committee of their accountants, in preparing a uniform system of accounts. They did so; and the system thus prepared by them, after being approved by the board, was put in operation. The accounts of all the Massachusetts roads have since been kept in practical accordance with it.

This measure carried the Massachusetts method of dealing with the railroad question to its ultimate point of development under a state government. No greater degree of publicity was possible. The system was perfectly simple, but none the less logical and practical. It amounted to little more than the establishment of a permanent board of arbitration, acting without any of the formality, expense and delay of courts of law. On each question which came before it,—whether brought to its notice by means of a postal card or through the action of a city government,—this board was to make an investigation. If wrongs and grievances were made to appear, and no measure of redress could be secured, the appeal was to the courts or the legislature, the board still being the motive force. Thus on all questions, not strictly legal, arising out of the relations of the railroad corporations,—whether among themselves, with the community as a whole or with individuals,—a body of experts, supposed to be skilled, was provided, who were clothed with full inquisitorial powers and whose duty it was, whether moved thereto by facts within their own knowledge or brought to their knowledge through the intervention of others, to investigate the doings or condition of the corporations, and to lay the resulting facts in detail before the public. Without remedial or corrective power themselves, behind them stood the legislature and the judiciary ready to be brought into play should any corporation evince an unreasonable spirit of persistence, when once clearly shown to be in the wrong.

The policy thus described would seem to have worked sufficiently well in Massachusetts. The commission has certainly succeeded in sustaining itself, for, while at every session the legislature has conferred upon it new powers, always in the same direction, the railroad corporations have never appeared in opposition to it as a body. The particular measures recommended by it have not, of course, always been looked upon with favor by the corporations, nor adopted by the legislature. This also has been fortunate, as the opposition and consequent delays thus encountered have given the commissioners time to reconsider many of the conclusions which they had reached and, if need be, to revise them. This they have frequently done. That a commission organized on a like basis in any of the western states between the years 1870 and 1875, could have accomplished the work there to be done is, to say the least, improbable. It could have commanded the confidence of neither side, and would have been listened to by neither. The issue then and there presented had to be fought out with other weapons than written reports. Now that it has been fought out and decided, the question presents itself in a different aspect. But it has been denied that a similar policy would, even under the conditions which now exist, succeed in the more western states, on the ground that the railroad corporations of that section are not so sensitive to the public opinion about them as those of Massachusetts. They are not owned in the West, and the absentee owner is currently supposed to care nothing for the West, its interests, feelings or sentiments. As to the local management, that is in the hands of salaried subordinates who are neither expected nor disposed to serve two masters. Every one at all acquainted with the real facts in the case knows perfectly well to how little weight this line of reasoning is entitled. The conditions stated are, on the contrary, exactly those in which such a machinery as that in use in Massachusetts would be peculiarly useful and effective. So far from being insensible to public opinion in the West, the eastern owner of western railroad securities is in fact peculiarly sensitive to it. The rule that the remote and unknown is always formidable applies in his case as well as in most others, and heretofore it has been by no means one of the least difficulties of the western railroad situation that no machinery has existed through which the voice of complaint could be carried over the head of the local manager directly home to the foreign proprietor. It is rarely indeed that satisfactory results are brought about through any dealings with subordinates. Nor is this remark true of railroads alone. Where any good cause of complaint exists, the local managers are usually more or less responsible for it. Men of routine, they as a rule can see only their own side of any question; above all, they rarely know when to yield, and are apt to resent interference as if it were an insult. The true way would be to go unceremoniously over them. The owner is the man. In the case of the western railroads, however, the foreign owner has naturally looked upon the Granger commissioner as the willing agent of demagogues and communists,—their mere attorney, bound to prosecute and annoy, right or wrong;—on the other hand, the commissioner has not hesitated to give his opinion of the foreign owner as a "robber baron," a "bloated bondholder" and "a money shark." To one who has had an opportunity to look behind the scenes on either side a wilder case of mutual misapprehension could not well have been imagined. A great advance towards a better condition of affairs in this respect has, however, been secured during the last year, through the action of the Iowa legislature in repealing the so-called "Potter" law, and substituting for it a commission practically organized on the Massachusetts plan. It will only remain for those who compose that commission to fairly try the effect of intelligent public discussion as a substitute for ignorant force. That the experiment should now be tried by them, and made to succeed, is of the utmost importance;

for if it does succeed the whole movement in the West will be advanced by one entire stage. The decision of the Supreme Court in the Granger cases having finally settled the legal relations of the parties, the discussions before this board and its consequent action may gradually establish them on a friendly and intelligible basis.

No comprehensive solution of the American railroad problem need, however, now or at any time, be anticipated from action of the government. The statesman, no matter how sagacious he may be, can but build with the materials he finds ready for his hand. He cannot call things into existence nor, indeed, can he even greatly hasten their growth. If he is to succeed, he must have the conditions necessary to success. So far as the railroad system of this country is concerned in its relations to the government, everything is as yet clearly in the formative condition. Nothing is ripe. That system is now, with far greater force and activity than ever before, itself shaping all the social, political and economical conditions which surround it. The final result is probably yet quite remote, and will be reached only by degrees. When it comes, also, it will assuredly work itself out; probably in a very commonplace way. The development will then unquestionably be found to have been correspondent; that is, consciously or unconsciously, the government on one side and the railroad system on the other will have worked towards each other. Whether travelling on lines nearly parallel, or which seem gently to converge or to sharply diverge, or even to run counter to each other, we may rest assured that, whether we see it or not, they are steadily in the United States, as in France, England and Germany, doing this now.

Hitherto the attempt has been to show how far the process of governmental development has as yet gone in America towards this common ground. It is not much and can be briefly summarized. So far as legislation, pure and simple, is concerned, no progress at all has been made. The laws intended to abstractly solve the difficulties presented have been mere copies, whether intentional or not, of similar acts long since passed elsewhere, and the utter futility of which is denied by no one. Passing on to the more positive results, the essential fact that railroad corporations are amenable to the legislative power has been completely established in the West; while in the East the influence of publicity and the resulting force of public opinion as a power adequate to all necessary control of the railroad corporations have been tested to a certain extent. The real issue, however, has not yet been touched; for all that has yet taken place is little more than the skirmishing which precedes a decisive engagement. Hitherto the question has been confined within state limits; but the problem in its full magnitude and complexity is co-extensive with that continental field in which Congress alone has " power to regulate commerce between the states." It is there, as the result of an uncontrolled and an as yet uncontrollable competition, that those harsh inequities manifest themselves which have hitherto baffled all attempts at regulation.

It is now necessary to turn from the side of the question which relates to the government, and to consider that which relates to the railroad corporations. It will be found that they also have made progress within the last few years. There is no occasion to go back beyond 1873. At that time the unnaturally rapid construction which had for ten years been going on produced its result. A general collapse took place. In that collapse the railroad interest from the first suffered more severely than any other, and a vast number of corporations were forced into bankruptcy, either because the country did not afford a sufficient business to support them or because such business as it did afford was fought over and competed for until it ceased to be worth possessing. Bankruptcy, again, became merely the process through which absorption was carried on; but it was a terribly exhausting process. It was competition run mad. So long as the struggle was confined to solvent roads, or to roads which had not yet resigned themselves to a condition of chronic insolvency, something might be predicated in regard to it. There was a point at which the owners of the railroads would cease to be willing to do business in a manner which seemed likely to result only in their inevitable ruin. The moment that point was reached and the conviction was fairly forced upon the minds of the contending parties that a conflict further prolonged would lead to this result and that shortly, then the moment for an agreement or for a combination had arrived. They invariably came together and sought to save themselves at the expense of the community. In other words, there was always a point, so long as solvent roads only were concerned, at which competition naturally and quietly resulted in combination. This, however, was true only of solvent corporations. But the effect of the crisis of 1873 was sharply to divide the railroad system of the whole country, and more particularly the railroad system of the West, into two classes: the solvent roads and the insolvent roads. The trunk lines mainly belonged to the former class, and the latter class comprised certain of the trunk lines and many, if indeed in the West not a majority, of what are known as the cross lines and the side lines.

Between the solvent roads and the roads thus bankrupt a new form of competition then developed itself. The bankrupt roads were operated not for profit, apparently, but to secure business; business at any price. If it was paying business, so much the better; if, however, the business would not pay, it was better than no business at all. Accordingly, the position of the solvent lines soon became almost untenable. They found themselves forced to decide whether to lose their business entirely and to see it pass away from them to rival lines, or to retain that business by doing it at a dead loss, which seemed inevitably to endanger their ultimate solvency also. Such competition as this could not terminate in the usual combination. The difficulty was exceptional, and unless some new method of solution could be devised, must be left to solve itself.

Accordingly, in the summer of 1873, those managing the principal through lines running east and west met together in conference. Commodore Vanderbilt was then passing the vacation time after his usual manner at Saratoga. To Saratoga the representatives of the other lines accordingly found their way, and there took place a consultation which became subsequently famous as the "Saratoga Conference." That conference resulted, it is true, only in a scheme which soon proved abortive; nevertheless it was deserving of all the temporary notoriety it achieved, for it will probably be found to have marked an era in the history of American railroad development; the era of what may, perhaps, be known as the Trunk-line Protectorate. There were five rival through routes. Chief among them was the New York Central. North of the New York Central was the Grand Trunk, the through route of Canada. South of it lay three other competing lines: the Erie, the Pennsylvania, and the Baltimore & Ohio. Of those lines three only, with their connections, were represented at the Saratoga conference, or agreed to its conclusions. These were the New York Central, the Erie, and the Pennsylvania. At the time, the results of the Saratoga conference excited alarm and popular clamor throughout the country. It was looked upon as a movement against public policy, and the plan for operating the combined roads which resulted from its deliberations was denounced as one which, if successfully carried out, must necessarily result in the destruction of all competition for carriage between the sea-board and the West, and as consequently turning over to a band of heartless monopolists the vital work of transporting the cereals of the interior to their market. The cry of the "railroad kings" and "railroad extortioners" was at once raised from almost every quarter. Meanwhile this clamor, like most popular clamors, had little real cause. The essential principle of the Saratoga combination lay in fact merely in the substitution of an open and responsible organization for a secret and irresponsible one, which had for years been in existence. To thoughtful and reflecting men it seemed very questionable whether, after all, such a change was not directly to the advantage of the community; even more to the advantage of the community indeed than of the railroad corporations. That the whole business of transportation between the West and the sea-board, and the prices which should be charged for doing it, had long been performed under common tariffs binding on all the roads represented at Saratoga, and made by their agents at stated times, was a matter of public notoriety. The newspapers had for years contained, among other regular news items, the reports of the meetings of these freight agents of the different corporations for the purpose of effecting these common tariffs, just as they had contained reports of the doings of the state legislatures or of Congress. That such meetings should have been held and such common tariffs prepared and published, was obviously a matter of mere necessity to the railroads. It would have been utterly impossible for them to live under the pressure of a war of rates knowing no limitation, —a war in which freight of every description should be transported long distances absolutely for nothing. There was a time, for instance, when cattle were brought over the competing roads in New York at a dollar a car. Such competition as this plainly opened the widest and shortest way to insolvency, and it was to avoid it that the conventions of freight-agents met. There was no secrecy about their proceedings. The tariffs arranged by them were published in the papers. They took effect at stated periods, and they were subject to modifications at other periods. There was no more concealment about them, if indeed so much, as there was about the regular local tariffs in operation on the several roads represented. The only difference between the local and the through tariffs was that, whereas the former were fixed and rarely changed, the latter were subject to sudden and violent fluctuations. These fluctuations were known as railroad wars, and to these it was proposed to put a stop through the machinery devised in the Saratoga conference. It was not intended as the result of that conference to, as it is called, "pool" the profits of the different lines which were parties to it. On the contrary, each line was to be left free to procure all the business that it could, and charge the agreed-upon rates therefor, and to keep to itself all the profits that it could realize from it. There was nothing which looked to a common-purse arrangement. The effort was solely to do away with wars of rates through the agency of arbitration. In place of the "Rob Roy plan" of leaving each company to assert its own rights and to maintain them if it was able, a central board was organized, the duty of which was to establish rules and tariffs which should be binding upon the various companies, and this central board it was intended should be clothed with sufficient powers to hold the companies firmly. It was an attempt to substitute arbitration among railroads for a condition of perpetual warfare; consequently, though the roads through this board secured a much closer combination than had ever before been effected, yet, from the very fact of their so doing, they also concentrated responsibility upon the board and consequently upon themselves. The board of arbitration was their representative. It acted openly and publicly, before the whole country. It established rates, and it was responsible to the country and to public opinion for the rates thus established. Upon it, therefore, the whole force of public opinion could, at any time, be brought to bear, in place of being dissipated as before among a number of wholly irresponsible subordinate agencies. Apparently, therefore, to any one who looked below the mere surface of things,—to any one who was not led astray by empty cries against railroad kings, and by the equally empty denunciation of monopolies, the Saratoga conference had resulted in no insignificant public benefit. It had substituted the responsible for the irresponsible; publicity for secrecy; it seemed, at last, to promise to bring the railroads together under one head, and that a directly accountable head.

Obviously, the adhesion of all the trunk lines was essential to the success of this experiment. The position would not be greatly altered from what it had been before, if, while the three central through lines between the West and the sea-board had effected a combination, they were yet flanked, as it were, on the one side and the other, by lines not parties to the arrangement; by the Grand Trunk upon the north, and by the Baltimore & Ohio on the south. This proved to be the fact. At the time of the conference, Mr. Garrett, the president of the Baltimore & Ohio, was absent in Europe. Immediately on his return, ostensibly to pay him a visit of compliment but in reality to induce him to give in his adhesion to the new arrangement, the representatives of the other lines paid a visit to Baltimore. It soon became apparent that trouble was impending. Mr. Garrett declined to surrender what he called the independent policy of the company which he represented. He professed the utmost willingness to agree on its behalf to adhere to the rates established by the combined lines, but he refused to subject it to the jurisdiction of the board of arbitration. He sought, in fact, to avoid all entangling alliances, and to keep the Baltimore & Ohio in a position of absolute independence, to do what it pleased in view of the local interests which it had always been its policy to foster. The representatives of the three central lines returned, therefore, from Baltimore in no good humor. Nor were their apprehensions of impending trouble unfounded. Hardly was the board of arbitration under the Saratoga conference organized, when a bitter railroad war arose between the lines which they represented and their southern neighbor. The more active hostilities were necessarily confined to the Pennsylvania road, which was brought immediately in contact with the Baltimore & Ohio. The war, though short, was very severe, and, for the time being, seemed to disorganize the railroad relations of half the country. It ended, as wars between solvent corporations always have ended and always must end, in an agreement. The Baltimore & Ohio became one of the combination of roads, upon the old footing of tariffs agreed upon in conferences of freight-agents. It retained its inde-

pendence. It was not subject to the jurisdiction,
or bound by the action, of any board of arbitration, and consequently the board became a useless
piece of lumber. Thus the one thing, practically,
which the furious struggle had resulted in was the
destruction of that which was best in the Saratoga
arrangement. The worst features of the old system of irresponsible combination were restored;
for with the board of arbitration the two great
principles of publicity and direct responsibility,
which that board necessarily represented, had also
disappeared; there remained nothing but a loose
understanding, such as it was, between four of the
five through routes, which was binding upon them
only so long as they saw fit to be bound by it.
Even this loose understanding, however, was not
a general one. The Grand Trunk of Canada refused to enter into it; and the Grand Trunk of Canada was not only thus a recusant road, but it also
so happened that it was bankrupt. This, for the
reasons already stated, sorely complicated the struggle. The combined and solvent roads were very
loath to enter into a war of rates with an insolvent through line, aided, as it necessarily was,
by the whole system of bankrupt western connections. Therefore railroad competition in the win-
ter of 1875 developed itself to its full extent, but
now with results which few had ever anticipated.
It, of course, led to discrimination, but for once
the places specially discriminated against were

the two great railroad centres of the country,— Chicago and New York. The Grand Trunk road led directly to neither of these cities. Consequently, the combined roads being unwilling to meet that line in a war of rates it was left at liberty to compete at points it did reach almost without restraint. Its rates, and those of the roads which connected with it, accordingly were marked down low enough to cause business to be turned away from the combined lines. This meant that business was diverted from Chicago and from New York, the centres which those lines especially connected. Meanwhile, though the Grand Trunk did not reach either Chicago or New York, it did through connecting roads reach the rival cities of Milwaukee and Boston. Hence it was that so long as that war of rates was suffered to continue, both New York and Chicago looked on, not without dismay, while the stream which flowed through their own channels seemed rapidly to be drying up, and that which flowed through the channels of their rivals was swollen beyond all precedent.

That such a condition of affairs should long be endured in silence was not to be expected. Accordingly the business communities of both cities soon began to bestir themselves, and the press of each to make itself heard. The course pursued in the two cases was almost diametrically opposite. In Chicago a committee of the Board of Trade was appointed to take the situation under advisement, and this committee in due time was delivered of a report. It was a singular document. The theory was advanced in it that what Chicago, the great converging point of all western railroads, yet needed was one more through line to the East; but this time it needed a hopelessly, permanently, bankrupt line in the hands of a perpetual receiver! In New York a far more sensible course was adopted. The merchants met together in conference, and a committee was appointed to wait upon the managers of the New York Central and to point out to them the damage which was being done to what must after all, under any circumstances, remain the natural terminus of their road. The duty of protecting their own best customer, which devolved upon those managing the line, scarcely needed to be dwelt upon. The meeting between the committee and the officials was a very friendly one. No complaint was made as to the rates then charged by the New York Central. These were freely acknowledged to be reasonable and sufficiently low. But the competing rates of the other line were lower. On this point there was no dispute, for the railroad officials freely admitted that the rates west from Boston were some fifty per cent. less than the rates at that same time from New York. It was not denied, either, that the condition of affairs necessarily resulted in great hardship, and must involve the destruction of many branches of New York business.

Under these circumstances, the Vanderbilts at once recognized and acknowledged the public duty which devolved upon them. They stated to the committee the circumstances under which they were placed, and promised that, at whatever cost, the interests of the city of New York should be protected.

A fierce railroad war now seemed impending. A bold announcement was immediately put forth that the New York Central was prepared to enter into the field of competition, even with its bankrupt rival, and that rates would be marked down to any point necessary for the protection of New York interests, however low that point might prove to be. Accordingly they were at once reduced some sixty per cent. It was obvious that events must take one of two courses. Either there must be a destructive war, in which the New York Central, as the solvent line, would suffer the most; or it must be made worth the while of those managing the Grand Trunk to enter the combination and retire from the struggle. Events moved rapidly. Scarcely were the newspapers filled with the rumors of war and with the loud notes of preparation for it, when they also announced that a conference of the competing parties was about to be held in the city of New York. It was held there. The usual discussion took place in public, which promised, apparently, to produce small results. The parties seemed to stand too far apart from each other. These things, however, are not generally arranged in public, or in the presence of newspaper reporters. While the representatives of connecting roads, east and west, were discussing and hopelessly differing, those representing the three corporations most immediately concerned withdrew to the parlor of a neighboring hotel. In an hour or two they separated, and the evening papers of New York for that day announced that, all differences between the competing lines having been adjusted, rates would at once be restored to a paying basis.

This combination was effected in December 1875, and was based upon a division of business; the longer and more indirect route being in fact allowed a share of the traffic, in order to buy it off. Matters were, however, arranged in that quarter only to have trouble break out elsewhere in an even more aggravated form. The combination of December, 1875, was, in fact, of even shorter duration than any of its numerous predecessors, for it lasted scarcely one month. Early in February it was broken, in consequence of a misunderstanding between the Erie and New York Central, and a new war of rates was begun on all east-bound through freights, under the influence of which they fell rapidly. This continued until March, when another meeting of the representatives of through lines was held, and renewed efforts were made to effect an understanding. These, however, resulted in nothing, except a brief postponement of an inevitable struggle. They wholly failed to touch the real root of the difficulty. This no longer lay in the old and chronic inability of the railroad officials to put any trust in each other's good faith, and rigidly to enforce upon their subordinates a scrupulous regard to agreements. The struggle had assumed a new and, to the railroad interests, far more dangerous form, that of a bitter rivalry between the great commercial cities of the seaboard. Baltimore and Philadelphia were not only asserting an ability to compete with New York as exporting points for western produce, but, owing to the thorough organization and perfect development of their great railroad lines and terminal facilities, they were demonstrating their power to do it. Ever since the opening of the Erie Canal in 1825, a monopoly of the business of exporting produce had been practically conceded to New York. Only within the last ten years has it been supposed that railroads could compete for the carriage of cheap and bulky articles with lake or even slack-water navigation. Rates, however, have continued to fall, until it has at last been demonstrated that under certain favorable conditions it is more advantageous at all seasons to forward nearly every description of merchandise by rail than at least by canal. Accordingly, the amount of agricultural products carried by rail from west to east, as compared with that carried by water, had gradually increased until, at the close of the year 1876 it amounted to more than half of the whole quantity moved. In 1873, the proportion was 29.8 per cent moved by rail, to 70.2 per cent by water; in 1874, it was 33 per cent by rail, to 67 per cent by water; in 1875, it was 41 per cent by rail to 59 per cent by water; and at last, in 1876, it was 52.6 per cent by rail to 47.4

by water. This transfer, also, had taken place notwithstanding the fact that during the years named the pressure of competition had forced down rates on wheat carried by lake and canal from Chicago to New York by more than one-half,—from 19.2 cents per bushel to 9.5. Lower than this they could not go at the canal tolls then in force, and at this rate the railroads were taking the traffic. Under these circumstances, it was inevitable that a wholly new phase of competition must be developed. Canal navigation was possible to New York alone; but when the traffic passed from the canal to the railroads, other cities possessed equal if not superior advantages. Accordingly, the struggle was no longer between the railroads leading to New York and the Erie Canal, but between railroads leading to different seaboard points. The monopoly of New York was threatened. Neither was the result of the impending struggle by any means so certain as long habit might induce many people to suppose. The prescriptive enjoyment of an undisputed monopoly has produced in the case of New York the usual results, and both railroads and business community of that place, confidently relying on long possession and natural advantages, have allowed abuses to creep in, or failed to supply improved facilities, until the handling of produce for export there has become most unnecessarily expensive. Meanwhile, the cities of Philadelphia and Baltimore, having great natural disadvantages to overcome, were naturally forced to husband every resource and make the most of every circumstance in their favor. All this they did with a degree of sagacity, foresight and success well deserving of careful study.

At bottom it was this changed relative position of Baltimore and Philadelphia towards New York which, during the early months of 1876, was gradually driving the great lines into a fiercer and more destructive war of rates than had ever been known before. New York, and consequently the main railroad line leading to it, began for the first time to realize that its easy supremacy no longer existed, and that in the struggle of competition it had no advantages to waste. Theretofore it had always been the practice on shipments from western points to the seaboard to take into consideration the distances of the several cities from the point of starting. A concession had always been allowed in favor of the southern points of shipment, under which originally the rate to Boston had been five cents per hundred more than to New York, that to New York five cents more than to Philadelphia, and that to Philadelphia five cents more than to Baltimore. These differences had subsequently been modified until, for some time previous to March, 1875, on all export merchandise rates to Boston and New York were equal, while those to Philadelphia and Baltimore, though equal to each other, were five cents less than the New York-Boston rate. As the sense of pressure from the competition of the more southern thoroughfares increased, however, the New York interests began to realize that this arbitrary rate placed them under a too heavy disadvantage. Accordingly, a new adjustment of rates was effected on a different principle. A differential tariff was agreed upon, based on distance, under which, taking Chicago as a fixed point, and the rate from that city to New York as the standard, a reduction from it of 10 per cent was allowed in favor of Philadelphia, and one of 12.5 per cent, in favor of Baltimore. This adjustment did not affect the Boston rate, which remained as before.

The results of the new system soon began to show themselves and it was apparent that the New York Central had been over-reached. The receipts of western produce increased immensely at Philadelphia and at Baltimore, indicating an alarming diversion of the export trade from New York; for the difference in rates between the ports was not infrequently almost equal to the entire ocean freight to Europe. When those controlling the New York Central became fully awake to this fact, and when they also realized the pressure in the way of equal competition under any circumstances which Baltimore and Philadelphia, with all their perfect facilities for handling through business, could now bring to bear upon them, it naturally occurred to them that the time had come for refusing longer to concede a differential rate in favor of those who seemed in no respect less advantageously placed than themselves. In order, however, to assume a consistent position on this subject, it became necessary for the Central road to extend the principle beyond New York, and to claim a uniform rate from the interior to all the seaboard points. This principle it was perfectly obvious that the southern or shorter routes could not concede, as with their longer ocean route it was tantamount to the surrender of that export trade a share in which they had made such prodigious efforts to secure. A full trial of strength thus became inevitable.

The struggle did not, however, break out in the first place between those who subsequently became the principal parties to it. On the contrary, all through the month of March and the early part of April, 1876, conferences were held and strenuous efforts made to hold the through lines to an understanding among themselves. At the last of these, on the 4th of April, the New York Central represented that it was under the necessity of meeting the competition of the Grand Trunk in New England, and to this those representing the other lines assented upon the understanding that the struggle was to be a local one, and was not to extend to New York, nor to divert business from that city. In the course, however, of a very few days, it became apparent that the contest could not be thus restricted, and as the result of a further conference on the 18th of April, at which a number of complaints were presented, the New York Central finally gave notice of the complete abandonment of all agreements, and almost immediately a general war of rates began. Between the 3d of May and the 14th of June, the fare between Boston and Chicago over the New York Central fell from $25.85 to $14, and that over the Grand Trunk from $23.85 to $12.50; while, as respects freights, the rates between Boston and Chicago on articles of the first class fell from 75 cents per hundred pounds to 20 cents, and that on agricultural products from Chicago to New York fell from 50 cents per hundred to 18 cents. These, also, were the published rates, while innumerable special contracts on terms far more favorable to shippers were made wherever business was competed for. Shippers whose patronage was really worth having were, in fact, in a position to dictate their own terms; and they did it. For six months the spectacle was witnessed of railroads hauling merchandise 1,013 miles east for $3.60 per ton, and the same distance west for $2.80 per ton,—in the one case at the rate of 3.5 mills per ton per mile, and in the other at the rate of 2.8 mills; a result which made sober and reasonable the most extravagant predictions which the advocates of cheap transportation had ever ventured to utter.

No sooner was the struggle fairly developed than the true issue was boldly avowed by the New York Central,—it being to restore the commercial supremacy of New York, imperilled by the rapid development of southern rivals. The principles of transportation involved were sufficiently simple. It was conceded on all sides that in the case of rival or competing lines between any two given points, as Chicago and New York, the shorter or more direct route had the right, as it was termed, to establish the rate; that is, it fixed a rate, and the longer routes were obliged to meet it, regardless of their own greater mileage; the principle of charging so much per ton per mile being, for obvious reasons, inapplicable. Where, however, lines terminated at different though competing centres, it was maintained that the principle of mileage should apply,—that there was no reason, for instance, why Baltimore should not enjoy, as compared with Portland or Boston, the full advantage of its geographical situation. While the managers of the Baltimore & Ohio and the Pennsylvania roads insisted,. therefore, on the differential allowance, those of the New York Central met them by fixing rates at so low a point that the differential allowance, when insisted upon, could not amount to enough to influence the course of traffic. Before the substitution of steel rails for iron, the roads could not possibly have endured the test of carrying on a season's business at the rates which ensued. Some idea may be realized of the wonderful economy which has been attained in the movement of merchandise, from the fact that as a regular thing a ton in weight was moved four hundred and fifty miles from Buffalo to New York for $1.50, whereas in the early part of the century it would have cost $100.

Having practically lasted for over eight months of the centennial year, thus enabling thousands of visitors to attend the Philadelphia Exposition at rates of fare never before dreamed of as possible, the struggle came to a close in December. In its results it was not decisive, for nothing but a general amalgamation of all the trunk lines into one huge consolidated company could have been decisive; and for that neither the corporations themselves nor the public mind were yet ripe. Though not decisive, however, the results were most significant. Without having secured, or probably ever having expected to secure, everything which it had claimed, the New York Central interest emerged from the conflict distinctly in the ascendant. It had stood the strain better than any of its rivals; its dividends were not diminished, and its balance sheet exhibited fewer items of mystery. Meanwhile the Baltimore & Ohio had abandoned all pretensions to independence, and was quite ready to enter into alliances, no matter how entangling; and the Pennsylvania road, having lost for the time being its ability to pay dividends, had lost with it all disposition to compete. The arrangement first arrived at was a very complicated one, in which an elaborate system of rebates to equalize ocean freights was a prominent feature, and which was apparently designed to secure a sort of rough equality among the several exporting cities; while, so far as the local traffic was concerned, a slight advantage was conceded to New

York in the northern portions of the great district in which through their connections all the trunk lines compete, and a somewhat larger advantage was given to Baltimore in the southern portions of that district. The fundamental idea of this arrangement was, however, reasonable and just, and has not since been abandoned. So far as the much sought for export trade was concerned, the entire cost of carriage from the gathering point in the West to the foreign market was taken into account, and the railroad rates to the several seaboard cities were so fixed as to give a decided superiority to no one of them. So far as was possible, they were placed upon a footing of equality, and left to improve their advantages or to overcome their disadvantages as best they could.

It soon, however, became apparent that something more than a vague understanding was necessary. Even Mr. Garrett had been forced to recognize the fact that the time was past in which railroads could insist upon their right to take all they could get, and to keep all they took. To deal with the great problem of competitive business a system, indeed, was wanted, but, even more than a system, a man.

Two years had now elapsed since the failure of the experiment devised at Saratoga, and little confidence was felt in any recourse to a board of extemporaneously devised commissioners or arbitrators who, however personally respectable they might be, would certainly be sustained by no external power, and would probably enjoy no particular degree of moral confidence. Meanwhile, as is always the case where a real want exists, the Saratoga failure had been followed by new experiments in the same direction elsewhere. To one of these experiments it is now necessary to recur. Uncontrollable and ruinous railroad competition was not at all confined to any single section of the United States, or to any particular system of railroads. It had been felt at the South even more than at the North and West, and the corporations there were in no condition to bear through any long period a heavy strain on their resources. It was in the South, accordingly, that the next attempt was made to reduce the railroad chaos to some degree of order. In principle this experiment was very like that which resulted from the Saratoga conference, but its working details had been much more carefully thought out. It assumed a definite shape under the name of the Southern Railway and Steamship Association.

The representatives of some thirty independent railroad and steamship companies had met at Atlanta, Georgia, in September, 1875, and regularly associated themselves. A formal constitution, setting forth both the objects the association was designed to secure and the means through which it was proposed to secure them, was then agreed upon and signed. Under this constitution the associated companies proposed to transact that portion, and that portion only, of their business in which they might be jointly concerned, and to the proper conduct of which constant negotiations and even coöperation were necessary. A central bureau was provided for, which was in fact a species of clearing-house. A single official, with the style of general commissioner, was to preside over this bureau. The necessity of transacting business through the clumsy agency of conventions was thus obviated, and, as all matters in dispute had to pass through the hands of an experienced and impartial officer, that personal contact between incompetent and irritated subordinates which is the cause of at least one-half of the railroad wars became wholly unnecessary. The general commissioner was intended to be the common executive officer of the association. As all negotiations were to be carried on through him, every difficulty as it arose necessarily came under his eye, enabling him to prevent many complications by judiciously acting as adviser and mediator. If, however, harmonious coöperation could not be preserved in this way, it then became the duty of the general commissioner, as umpire, to judicially decide questions at issue between the members of the association, though his decisions were at all times subject to appeal to a board of mutually appointed arbitrators. The next duty of the general commissioner was to see that all agreements entered into, and all his own decisions or those of the boards of arbitration, were fully and honestly carried out. In this respect, of course, he, like the Saratoga commissioners, could bring no legal power to bear on a recusant. Yet, though the force he could exercise was in main a moral one only, he was not confined to that. He could, in case of need, declare a partial or even a general war of rates, and the combined force of the association being thus wielded by one hand, it was in a position to practically enforce a policy, and, what was more, in doing so to expend only that amount of strength necessary to accomplish the end in view. Neither could the withdrawal from it of any one member, nor indeed of a number of them, dissolve the association. For, in spite of such withdrawal, the clearing-house and the agency for the transaction of joint business still remained in the service of those which were left. As other companies could also at any time join the association, the system admitted of indefinite expansion, and, indeed, could with mere changes in detail be made to include the entire railroad system of the continent, much as the similar German association includes all the railroads of Central Europe.

Next to the outside pressure which causes those managing the individual lines to yield something of their independence for the sake of order, the success of any such combination as this depends almost exclusively upon the ability, temper, and skill of the general commissioner, and upon the degree with which he is able to inspire respect and confidence in the minds of the associates. In this particular the southern association was fortunate. Colonel Albert Fink, for a number of years connected with the Louisville

& Nashville Railroad Company as its Vice President and General Manager, in which position he had displayed qualities especially fitting him for such a post as he was now chosen to fill, became its general commissioner. He, if any man, could be expected to carry the plan of the association into successful working, for it was he who had devised and matured it. The experiment was thus tried under the most favorable auspices, and its success, though not complete, had in 1877 been sufficiently marked to attract throughout the country the attention of more observant men interested in questions of railroad development.

Determined to go a step further towards an effective combination than they had ever gone before, those representing the northern trunk lines now summoned Colonel Fink to their assistance, and appointed him as commissioner to act under an executive committee of their number in the adjustment of all questions which might arise among them. The object of the new combination was at first simple and comparatively feasible. All the trunk lines whether they terminated there or not, had offices in New York, and were in steady competition for the merchandise shipments to the West from that place. Consequently not only was the value of the business destroyed to the railroads, but a steady discrimination existed in favor of New York as against other and especially more interior points. The combination of which Colonel Fink was the executive officer was intended, therefore, to regulate rates on all shipments by rail from the city and suburbs of New York to the West. The business was divided among the four trunk lines, on the basis of an apportionment of tonnage and not an apportionment of earnings. Daily returns of all shipments from New York to the West were made to the commissioner, and thirty-three per cent of this business was allotted each to the New York Central and the Erie, twenty-five per cent to the Pennsylvania, and nine per. cent to the Baltimore & Ohio. Any road doing more than its allotted share of business had, under the commissioner's instructions, to turn the excess over to the road doing less than its allotted share, and to pay full rates for it.

The operations of this organization were unnecessarily complex and cumbersome, owing to the fact that the roads concerned in it were unwilling to surrender their identity so far as to issue only common bills of lading from common offices. They insisted on issuing their own bills specifying the line by which shipment was made, and thereby incurring legal obligations. If, then, any company was found to have received and receipted for shipments in excess of its proportion, that excess had to be carted off to some other line, and reached its destination in a way different from that designated in the bill of lading. This method of dividing the business was, therefore, not only cumbersome, but it involved very serious legal questions which at once suggest themselves. It was due to the old jealousy. Each company, but the New York Central especially, wished to retain its own business and its organization for getting it. They were not prepared, so to speak, to burn their bridges, even though by so doing a great economy would have been effected, by doing away with an army of agents, and all questions of legal responsibility avoided.

Nevertheless in the hands of Colonel Fink, clumsy as its mode of operation was, the combination held together and a division of the business was effected. Its operation was peculiar, in that it was limited to the city of New York and its immediate vicinity. At other central points, such as Boston, Philadelphia and Baltimore, the local business was left to take whatever course it chose, the several lines merely agreeing to sustain rates. Should, however, the rates not be sustained at these local points, this fact did not necessarily involve a disruption of the New York combination. The west-bound business would still be distributed among the lines in the method already described, though the rates charged would, in certain events, have to be reduced to prevent a diversion of business from New York to the point of local competition, wherever that might be. This contingency, in fact, speedily arose through the action of the Grand Trunk line which, not reaching New York, proceeded to compete at Boston. A diversion of business was accordingly effected to such a degree that New York merchandise during the latter part of 1877 began to move to the West by way of Boston. This naturally occasioned loud complaint in the former city, and threatened through the mere force of public sentiment to either destroy the combination or to force it into a general war of rates which would have affected equally all the seaboard cities. The danger, however, was for the time being removed, and the acquiescence of the Grand Trunk secured in the usual way.

Encouraged by the success with which, under the lead of their new commissioner, they had effected a division of the business from the sea-board to the interior, the representatives of the trunk lines now turned their attention to the current which flowed in the other direction,—from the West to the East. Experience had shown that the problem of controlling the west-bound shipments was one sufficiently difficult to solve; it was, however, simplicity itself compared with that which presented itself in connection with east-bound shipments. These, in the first place, as is well known, constitute the vast bulk of the entire movement. Consisting chiefly of agricultural products, they require the use of an immense mass of rolling stock at certain seasons of the year, and of comparatively little at other seasons. The movement then goes on from an immense number of points, and through a great many channels, some natural and others artificial, but all competing. The consequence is that every conceivable agent of competition is brought into play. The utmost latitude for the uncontrolled action of these competing forces is further secured through the fundamental principle of the whole pro-rating railroad system, that the agent of the connecting line at the shipping terminus makes the rate, which the intermediate lines and the line at the point of delivery accept. In other words, the rates on east-bound freight at the western points of shipment were made by the well-nigh innumerable agents of multifarious and often irresponsible companies; and all that the intermediate and eastern companies had to do with the matter was to haul and deliver the goods, receiving their proportion of the freight-money for so doing, according to the distance hauled. Under these conditions, if any, an uncontrollable competition was naturally to be expected. No combination to control the rates at which the business should be done seemed practicable. Nevertheless the attempt was made on the basis of Colonel Fink's New York combination. In February and March, 1878, a succession of meetings were held, which were attended by the representatives of no less than forty different corporations operating some 25,000 miles of road, or about one-third of the whole system of the United States. After some preliminary difficulties had been disposed of a plan was adopted. Seven points of general railroad convergence at the West were selected. The representatives of the roads running east from these points met and agreed upon a division of the tonnage through and from such points, which was to hold good for a period of three months, to try the experiment. The rate from Chicago to New York was then fixed upon as the basis upon which rates from all other points were to be computed. For the rest the machinery was precisely the same as that already described in the case of the New York combination. Daily returns were made from each of the pooling centres to a chief commissioner, whose office was at Chicago, and a corresponding division of tonnage was effected through him. All questions which arose were referred to him for adjustment, and consequently the rate established for the time being at Chicago regulated the rates charged all through the great interior field of railroad competition. The difficulty with this arrangement, from an interior point of view, was, of course, its lack of any cohesive force. It depended on the good will and coöperation of too many persons. Nevertheless, an executory force, and a powerful one also, was not wanting. It was furnished by the trunk lines. Their position enabled them, if they saw fit, to give a very effective support to any order of the western commissioner. This arose from the fact that they are the stems to which the numerous western connections are branches, and everything which the latter collect has to be delivered over to some one of the five trunk lines to reach a seaboard destination. If, therefore, acting in support of the western combination, the trunk line combination had absolutely refused to connect with any road which failed to maintain rates, except on arbitrary terms, the result must have been that such road, shut up in its own territory, would, soon or late, have been forced to submit.

The three months during which it had been agreed that this plan of apportionment should be tried expired in June, 1878. At a meeting of those representing the roads interested in it, held in New York, it was then pronounced a failure, and abandoned; the commissioner himself admitting that he had been powerless either to enforce rates or to punish those who had defiantly violated the compact. Whether the cause of failure was due to the absence of the qualities essential to success in the commissioner himself, did not appear; but it was apparent that the trunk line combination had not seen fit to exert its power to further continue the experiment at that time. It allowed the western connections to begin another war among themselves, simply fixing its own share of the joint rate, which was sufficiently large to leave the combatants a somewhat beggarly margin to fight over. Nevertheless, the idea of a trunk-line Protectorate had been developed, and its feasibility proved. The direction of the next attempt is thus very clearly indicated.

The combinations which have been described are representative. Capital is trying to protect itself; and will succeed in doing it. The stress of competition has been too great, and in its own way is resulting in combination. It is not necessary to multiply examples. If it were they could be found in any direction, for during the last two years, there is almost no section of the country or branch of trade which has not been " pooled." The " Omaha pool," the " Colorado pool," the " Southwestern Rate Association," the " Southern Railway and Steamship' Association," are examples of these combinations working over sections of territory ;—the " Coal Combination," the " Chicago Cattle pool," the " Oil pool," are examples of their working over branches of business. They are based upon the same fundamental principles ; whether a division of business, or a division of the money receipts from business. They are directed to the same end, the control of competition. They differ only in details. The effect of these combinations and their evident tendency to development remain to be considered.

It is, of course, unnecessary to say that all combinations of the character of those which have been described are looked upon with much popular distrust, and are held to be against well established considerations of public policy. In the minds of the great majority, and not without reason, the idea of any industrial combination is closely connected with that of monopoly, and monopoly with extortion. In view of this fact, it is very pertinently asked,—Why should a railroad combination, avowedly intended to hold competition in check, if not to put an end to it, produce any result other than the natural and obvious one of raising prices?—Who or what is to protect the community against the extortions of these great corporations, should they cease to quarrel and compete among themselves?—In the first place, conceding all its evils and objectionable features,—its inherent wastefulness, its harsh inequities, its violent fluctuations,—the question still remains an open one, whether competition, as it has hitherto existed, is not after all at least for America the best and the final solution of the railroad problem. If, indeed, from this point of view, any railroad problem can be said to exist. The considerations which lead to the conclusion that it may be, have been very clearly stated in a recent public document:

"The evils incident to competition are sharply defined and incisive, but the benefits it affords are substantial and pervading. The beneficent law of supply and demand where it operates most freely may not secure systematic justice, and yet the whole world concedes that, so far as it is operative, it secures substantial justice. This is all that can be expected in the present condition of human affairs. Competition may not make all things even, but it affords a nearer approach to equitable dealing among men than any substitute which has yet been proposed for the natural laws of trade. The very *instability* of competition is the surest safeguard of the public interests. When competition ceases to be irresponsible, monopoly will step in, unless it be substituted by the autocratic rule of a combination sufficiently powerful to control all the transportation lines of the country. Any arbitrary rule in whatever manner formulated, or by whatever agency exercised, would prove to be an impotent substitute for the great beneficent law of competition in the irresponsibility and instability of which is embraced that conservatism which inheres in the untrammelled operations of natural forces.

"So intimately are the interests of transportation and of trade connected, that it is impossible to eliminate competition between the railroads without doing violence to commercial interests, and thereby working greater evils than those sought to be removed. . . .

"It has been supposed that in the contests between the trunk lines the strongest company or combination of companies invariably remains master of the field. It has, however, come to be almost an aphorism among railroad managers that the weakest line determines the rates. . . . It has also been supposed that the combination between the railroads of the country is yearly becoming closer and more powerful. The facts which have been hereinbefore adduced seem to indicate, however, that the extension of the railway system has tended to create new elements of competition, and to render the adjustment of through routes more difficult. Every trunk line has many interests outside of, and which cannot possibly be embraced within the terms of any combination with other trunk lines; and it has been found that sooner or later these collateral interests lead to the infraction of any conditions which the ingenuity of railroad managers has yet been able to devise. The difficulty appears to be that, heretofore, competing companies, in attempting to protect themselves, have failed to arrive at a clear understanding of the nature and limits of their mutual interests.

"As the promoters of the great railroad organizations connecting the West with the seaboard have pushed their lines westward, they have seen their control over 'through rates' gradually becoming weaker and weaker, and the idea has been suggested in the interests of the railroad companies, by able men, that the roads ought to invoke some external aid in order to maintain remunerative rates, or at least to avoid the necessity of at times carrying through freight at an absolute loss. Evidently, it would be detrimental to the public interests if the railroads of the country were to become crippled by their own excesses, but, in view of the beneficial results which have been realized from the regulating influence which competition has exerted over rates, in the great commerce between the West and the East, and between all important centres of trade which enjoy the advantages of two or more rival lines, the people will watch with favor the gradual extension of the railway system by which means the limits of the local or non-competitive traffic are being contracted, and the limits of competitive traffic enlarged.

" It appears probable as the facilities of transportation are more widely extended many injurious discriminations will disappear, and that the legitimate limits of the traffic of rival companies will become more clearly defined. It is also to be hoped that the various companies will, upon enlightened views of self-interest, formulate and acquiesce in such regulations with respect to their common interests as to prevent the occurrence of those sudden changes of rates which cause erratic diversions of traffic from one line to another; results which tend to destroy confidence not only in railroad securities, but in the value of the entire property of commercial cities. Such changes of rates tend to depress and not to advance commerce."*

The opening and closing considerations presented in this extract are somewhat at variance. A combination which will be equal to the difficult task of preventing " the occurrence of those sudden changes of rates which cause erratic diversions of traffic," and which would yet not be "sufficiently powerful to control all the transportation lines of the country,"—such a combination as this would be one difficult at least to devise. That the law of competition where it can have full play is a most beneficent one in its operations, few would care to deny. As respects the railroad system, however, the whole

* First Annual Report on the Internal Commerce of the United States, by Joseph Nimmo, Jr., pp. 88-91.

problem happens to be embraced in the question,— Can this law from the nature of the case there find full scope for its operations?—If it can, there is an end of the matter. There is then no problem at all to consider, for competition will settle the whole difficulty. But, on the other hand, does not competition in the case of the railroad system, necessarily, while working with excessive violence, work most unequally?—In fine, is not discrimination, somewhere and against some one, the logical and inevitable result of every uncontrolled railroad competition ? And is it not matter of experience, that the fiercer the competition grows, the harsher the discrimination becomes?—If this is so, can such results be classed as among the usual results of a " great beneficent law ? "—This is the whole question, if there is any question ; and it will not do to beg it. That " the weakest line determines the rates," is true; but it would also seem, from the experience of the past, that it will continue to determine them only so long as the stronger line does not see its advantage in absorbing it through lease, consolidation, or purchase. A railroad war which does not end in a consolidation of lines or in the absorption, through some process, of one line by another, is an indecisive war, and, judging by all recent experience, will almost inevitably be renewed, soon or late. That the absorption does not all take place at once, and that the combining process is not yet over, merely proves that all railroad, like other wars, are not decisive, and that great processes take time for their completion. So also it is true that the limits of local or non-competitive traffic are undergoing a steady contraction; but it is this very fact which has precipitated the recent destructive railroad wars. These, in their turn, have then rapidly led to schemes of combination between the trunk lines, while the smaller independent lines, exhausted by competition, have been forced to lose their identity in one or another of the contending systems. The consolidating and combining influences of the last ten years have, indeed, produced results not to be mistaken. There are, also, very significant indications that the railroad system of this country is now on the threshold of a most active and unprecedented development in that way. This subject, however, will more naturally be referred to in another connection.*

Nevertheless, the history of railroad development from the beginning has been little else than a constant succession of surprises. That which was most confidently predicted has rarely come to pass. It may well be, therefore, that in the present case those conditions which one class of observers regard as wholly transitional may by degrees be found to contain in them the elements of permanence; while, on the other hand, those indications which have led another class of observers to refer to events now taking place as "incidents in a phase of the process of de-

---

* See pp. 195–7.

velopment," may prove to have been mere surface movements, significant of nothing.

Meanwhile, the combinations of railroads, whether they are going to result in any thing more than they now are or not, do exist and do challenge the public consideration. That they have rapidly developed of late, is apparent. That they may develop yet more in the immediate future, is very probable. They involve in their continued existence the whole question of competition between railroads; and, whether they are to grow into a system or not, the results, so far as the public are concerned, which may naturally be expected to flow from them, either in their present or their possible form, are well worthy of discussion.

Contrary to the general and popular conviction, an increasing number of those who have given most thought to the subject, whether as railroad officers or simply from the general economical and political points of view, are disposed to conclude that, so far from being necessarily against public policy, a properly regulated combination of railroad companies, for the avowed purpose of controlling competition, might prove a most useful public agency. These persons contend that railroad competition, if it has not already done its work, will have done it at a time now by no means remote. An enormous developing force, during the period of construction, its importance will be much less in the later periods of more stable adjustment. Under these circumstances, and recognizing the fact that the period of organization is now succeeding that of construction, these persons are disposed to see in regulated combination the surest, if not, indeed, the only way of reaching a system in which the advantages of railroad competition may, so far as possible, be secured; and its abuses, such as waste, discrimination, instability, and bankruptcy, be greatly modified if not wholly gotten rid of. In conducting its traffic, they argue, each road or combination of roads is now a law unto itself. It may work in concert with other roads or combinations, or it may refuse to do so. It may make rates to one place, where it may think it for its interest that business should go, and may refuse to make them to another place where it is for its interest that business should not go. All this is essentially wrong. Yet the business community of America, from one end of the country to the other, has been from the beginning so thoroughly accustomed to the extreme instabilities of railroad competition, that it has wholly lost sight of what its own interest requires. What it needs is certainty,—a stable economy in transportation,—something that can be reckoned on in all business calculations,—a fixed quantity in the problem. This, of all results the most desirable, is now even looked upon with apprehension. There is an idea, the result of long habit, in the public mind, that, so far as transportation is concerned, prosperity is to be secured through a succession of temporary local advantages, —an unending cutting of rates. The idea of a great system of internal transportation at once reasonable, equitable, and certain,—permitting traffic to flow and interchanges to be made just how and where the interests of buyer and seller dictate,—never discriminating,—rarely, and then only slowly, fluctuating,— this is a conception very far removed from the reality, and it may well be doubted whether now it even commends itself when stated to the average man of business. He clings, on the contrary, to the burden of inequalities to which he is accustomed, and is inclined to doubt whether he could live without them. It is as if a mariner had become so habituated to a constant succession of squalls and simooms, that he questioned whether it would be possible to satisfactorily navigate a ship in trade-winds; especially if the trade-winds blew for all. Accordingly, equal rates, no matter how reasonable, the moment they are applied are looked upon by the favored points of competition, like Chicago or New York, as in some way an outrage. These points have become so accustomed to discrimination against others and in their own favor that they regard it as a species of vested interest. Their boards of trade call upon their legislatures to secure it to them. They have even gone further than this, and presented the somewhat ludicrous spectacle of modern communities claiming that their own want of enterprise and wasteful methods of doing business should be prevented from bringing forth their legitimate results through an unending railroad war. So possessed are they, indeed, with this idea, that it may safely be predicted that the principle of absolute freedom and strict impartiality in the management of the railroad system of America as a whole will only be reached as a result of long discussion and in the face of strenuous resistance. Meanwhile, until it can be reached, those composing the business public, as well as the stock-holding class, must reconcile themselves as best they can to frequent repetitions on an increasing scale of those wild fluctuations and ruinous discriminations which have just been described in detail in the account of the railroad complications since 1873. Nothing of the sort exists in any other portion of the world.

If, however, any approach is ever to be made towards that ideal state of affairs which has just been suggested, it can apparently be made only in one way. The abuses incident to unhealthy railroad competition must cease; and undoubtedly the first step towards getting rid of those abuses is to render the railroad system, throughout all its parts, amenable to some healthy control. The present competitive chaos must be reduced into something like obedience to law. Yet this apparently can only be effected when the system is changed into one orderly, confederated whole. To attempt to bring it about during an epoch of wars, and local pools, and conflicts for traffic, would be as futile as it would have been to enact a code of laws, unsupported by force, for the government of the Scotch Highland clans in the sixteenth century, or a parcel of native African tribes now. A confederation, or even a general combination among all the railroad corporations having some degree of binding force, might, therefore, as has been suggested, not improbably prove the first step in the direction of a better and more stable order of things. But to lead to any results at once permanent and good this confederation must, in three respects, differ radically from everything of the same sort which has hitherto preceded it: it must be legal; it must be public; it must be responsible.

Tried by this standard, it is safe to say that none of the combinations now existing are consistent either with sound views of public policy, or have in them the elements of permanence. They are, in the first place, secret combinations of *quasi* public agencies; in the next place, as respects the ends they have in view and the means they use to attain those ends, they are amenable to no law; and, finally, they are all in greater or less degree irresponsible even to public opinion. They fail, moreover, even to accomplish the one result which, if practically brought about, might justify their existence,—they do not afford to the community a reasonable and equable system of charges for carriage, permitting an unchecked flow of travel and commerce, the continuation of which may with safety be calculated upon. A local and makeshift character is apparent in them all, and is not ignored even by those who are parties to them.

Indeed, every disturbing element which has heretofore broken up other combinations is latent in those now existing; the individuality of organization, the distinct separation of traffic, the armies of local freight agents, the extending of connections. While the processes of pooling are going on and during the very periods of truce, there is not a single considerable line in any one of the combinations which is not always anxiously looking about to strengthen itself in case of an ever-expected renewal of war. Under these circumstances, they will probably last only so long as the recollection of recent losses and the costly pressure of the last railroad war is fresh in the minds of stockholders and officials. Indeed, it is safe to say that the greatest of all these combinations, that of the trunk lines, is held together only by the personal influence and force of character of one man, —its commissioner, Colonel Fink.

The difficulties which stand in the way of any permanent organization of a beneficial character can, moreover, hardly be exaggerated. They are indeed so numerous and so great that it is regarded as little short of visionary to suggest that they can ever be overcome. In the first place there is no one legally authorized to enforce the peace between the high contracting parties. Each reserves the right to construe any agreement to suit itself, and in the last resort to refuse obedience to the decision of any one. There is no court of common arbitrament with the sheriff's officer behind it. Without this, all railroad combinations in this country, where a division of territory is impracticable, will prove but temporary. Even were they, under certain conditions, practicable, they are now rendered more than precarious, owing to the fact that the whole complicated system under which through or competitive railroad business is done is curiously vicious and extravagant, and must be radically reformed as a preliminary to any final settlement. It now implies the presence of a vast army of subordinates whose very existence depends on that not being done, which those controlling the lines which feed them are continually trying to do. To realize the truth of this fact, it is but necessary for any person to walk down the leading business streets of any considerable town in the country. He will see that a great number of expensive offices bear the signs of railroad and of car and dispatch companies, and at them tickets can be purchased and rates of freight made which are binding on all the connecting lines. The rents, salaries and perquisites of the army of retainers who occupy these offices come out of the railroad corporations, and the interests of the retainers and the corporations are exactly antagonistic, —the first are always working to bring about railroad wars, in which business with them is brisk, while the last are always striving to effect combinations.

Beyond and behind all this, however, the railroad corporations of the United States have from the beginning enjoyed a sort of lawless independence. Corporations, like communities, accustomed to this, necessarily remain for a long time restive under any sense of control. They need constantly to feel that a policeman's eye is upon them, and that there is a stationhouse in the next street. No one or two great corporations have yet been developed with power sufficient to assume a coercive protectorate over the others and to compel obedience. The combination of the trunk lines and their recent action towards their connections in the West is the first approach yet made towards this result. But without the cohesive influence of some such protectorate there is in all voluntary combinations a natural tendency to anarchy. In the absence, therefore, of any compelling force to secure order and subordination, the mill of competition has got to keep on grinding for some time yet. Its work is not done. Indeed it will not be done until, through the process of its grinding, the great principle of the survival of the fittest is finally ground out.

This process is unlikely to prove a rapid one, for order is not easily established in any community which has been long in a state of anarchy. In such cases the demoralization becomes general; the tone of the individual deteriorates. This is what is now the matter with the railroad system in America. Lawlessness and violence among themselves, the continual effort of each member to protect itself and to secure the advantage over others, have, as they usually do, bred a general spirit of distrust, bad faith and cunning, until railroad officials have become hardly better than a race of horse-jockeys on a large scale. There are notable individual exceptions to this statement, but, taken as a whole, the tone among them is indisputably low. There is none of that steady confidence in each other, that easy good faith, that *esprit de corps*, upon which alone system and order can rest. On the contrary, the leading idea in the mind of the active railroad agent is that some one is always cheating him, or that he is never getting his share in something. If he enters into an agreement, his life is passed in watching the other parties to it, lest by some cunning device they keep it in form and break it in spirit. Peace is with him always a condition of semi-warfare; while honor for its own sake and good faith apart from self-interest are, in a business point of view, symptoms of youth and defective education. Under such circumstances, what is there but force upon which to build? It was the absence of the element of force which caused the failure of the Saratoga association, and probably will cause the failure of those which have succeeded it. Taken as a whole, the American railroad system is in much the same condition as Mexico and Spain are politically. In each case a Cæsar or a Napoleon is necessary. When, however, the time is ripe and the man comes, the course of affairs can even now be foreshadowed; for it is always pretty much the same. Instead of the wretched condition of chronic semi-warfare which now exists, there will be one decisive struggle, in which, from the beginning to the end, the fighting will be forced. There will be no patched-up truces, made only to be broken, for the object of that struggle will be the complete ruin of some one in the shortest possible time. Then will come the combination of a few who will be sufficiently powerful to restrain the many. The result, expressed in few words, would be a railroad federation under a protectorate. The united action of the great through lines is necessary to bring this about; and how to secure that action is now the problem. If the elder Vanderbilt were alive and in the full possession of his powers he probably would solve the difficulty in the way most natural to him. Meanwhile, although Commodore Vanderbilt is dead, there are very significant indications that his work is going on. His vast property, in the peculiar shape in which he left it and as it is now handled, seems to be little else than an accumulating fund devoted to bringing about a consolidation of railroad interests on the largest possible scale. The New York Central is the basis upon which this superstructure rests. The Vanderbilt interest in the property is so great that practically the earnings of this road, instead of being dissipated among innumerable stockholders, as is the case with the other trunk lines, are continually applied to securing the control of other and connecting lines,—first the Lake Shore, then the Canada Southern, then the Michigan Central. Scarcely during the last few years has one of these connections been absorbed before rumor has announced that operations have been begun upon another. The Erie, the Atlantic & Great Western and the Wabash are supposed to be the next in order. The precedent established by the father in buying up the Hudson River road in one lump in order to put a stop to its competition with the Harlem, is apparently being followed on a larger scale by the son.

While this is going on in the East under the Vanderbilt lead, two other and precisely similar "one-man" combinations are assuming shape,—the one in the central region of the country west of the Missouri, and the other on the Pacific slope. They, also, are built upon the principle of devoting the earnings to the development of the business, and not to the support of stockholders. Accordingly competition does not now exist within the sphere of influence of these combinations, and its existence is rapidly becoming impossible; for, as soon as it makes itself felt, the competing line is bought out of the way. In this way the Union Pacific combination now controls seven corporations owning and operating 3,000 miles of track in the heart of the continent,—an absolutely controlling interest;—while on the western coast the Pacific Central occupies an even more commanding position.

In presence of such a policy as is common to all these interests, backed by such resources, the "fighting" superintendent finds his proper level. His day is over. The period of indecisive railroad wars is drawing to a close. The development in these cases is moreover made with ready money;—that is, the earnings of the combinations are continually accumulated in the business. There is no toppling superstructure of debt. Those earnings aggregate millions a year. In view of these facts there would seem to be some ground for supposing, as was suggested in another connection, that the railroad system of this country is now on the threshold of a most active and unprecedented consolidating development, and that the question of the survival of the fittest among railroads may here be decided at a less remote day than is usually supposed.

But, however this may be, it is a question of the future and certainly has no immediate bearing on the existing combinations. Of these the only one which seems entitled to any thoughtful consideration is the Southern Railroad and Steamship Association. So far as the public is concerned, everything essential as a safeguard against abuse seems in the case of that association to be provided. It is a complete, but not a secret combination. It exists in the full light of publicity. The purposes for which it was organized are openly avowed, and its every transaction is, or may easily be made, matter of general observation. To secure this result it would only be necessary to give it legal recognition. By its originators it is confidently claimed that, if properly developed and recognized by legislation, it would afford a complete and practical solution of the American railroad problem. Whether it would or no, it is certainly a great advance on any other form of solution which has yet been suggested. It is at once far more philosophical, more practical, and more in consonance with American political usages and modes of thought. Indeed, it is not easy to point out any respect in which it might not fairly be accepted as the natural and logical outgrowth of American railroad development, as it has gone on up to this time. The difficulty with all of the many other solutions which have from time to time been suggested has been that they disregarded what had gone before; they were none of them in the nature of a logical sequence or natural outgrowth. Those who originated them sought to deal with a vast and complicated system as if it were so much plastic material, to be handled and shaped at will. Both the scheme for the state ownership of railroads, and the measures of Granger legislation failed and were abandoned, for this reason,—they were not sequences; and while the first could not stand the test of discussion, the last broke down in practice. The idea of regulating all railroads through the state ownership of one,—the Belgian system,—was some years ago brought forward, and urged on legislative attention by the Massachusetts commission as at least worthy of careful inquiry. Subsequently it was examined into more carefully in connection with a Massachusetts enterprise in which the state already held a controlling interest, and the practical difficulties and objections, both political and financial, which presented themselves were of such a nature that those engaged in the investigation, although one of them was a member of the commission and as such committed to the theory, unanimously found themselves compelled " by the logic of their investigations, regardless of preconceived theories," to another result.

Irresponsible and secret combinations among railroads always have existed, and, so long as the railroad system continues as it now is, they unquestionably always will exist. No law can make two corporations, any more than two individuals, actively undersell each other in any market if they do not wish to do so. But they can only cease doing so by agreeing in public or in private on a price, below which neither will sell. If they cannot do this publicly, they will assuredly do it secretly. This is what, with alternations of conflict, the railroad companies always have done in one way or another; and this is what they are now doing and must always continue to do, until a complete change of conditions is brought about. Against this practice, the moment it begins to assume any character of responsibility or permanence, statutes innumerable have been aimed, and clauses strictly interdicting it have of late been incorporated into several state constitutions. The experience of the last few years, if it has proved nothing else, has conclusively demonstrated how utterly impotent and futile such enactments and provisions necessarily are. Starting, then, from this point—accepting what is and what must continue to be—the fundamental idea of the Southern Steamship and Railroad Association is to legalize a practice which the law cannot prevent, and, by so doing to enable the railroads to confederate themselves in a manner which shall be at once both public and responsible. This is the railroad side of the question. The other side of the question—that of the public—admits of a statement equally clear. Its essential point, however, is that, through this process, and this process only, can the railroad system as an organized whole be brought face to face with any public and controlling force, whether of law or public opinion. Once let the railroad companies confederate in accordance with law, and the process through which this all-important confronting result would be brought about is apparent. The confederation would be a responsible one, with power to enforce its own decisions upon its own members. The principles upon which it could act, as a creature of the law, would be formulated in the law. It could compel obedience, but obedience only to legal decrees, and the question in each case would be whether the decree was legal. At exactly this point the machinery for state supervision would come into play in the form of a special tribunal; like those which have already been provided in England and France, or that now being matured in the Prussian parliament. The field of discussion before this tribunal would be commensurate with the whole subject of transportation by rail, including questions not only of law but of economy. Then, at last, the correct principles governing railroad traffic would be in course of rapid development. The essential features of what constitutes discrimination and extortion would gradually be formulated into rules, and the moment that is accomplished competition will work equitably. This result must follow. It must follow, from the fact that competition is now almost entirely local. That is, a competitive rate to or from one point, in no way necessarily affects rates to or from other points,—a local variation does not cause the whole schedule to move up or down. This is what makes discrimination. Could the system be confederated and equalized, however, such would not be the case. An established tariff, intended to be public and permanent, would then have to be fixed upon, just as it is to-day fixed upon in each of the local pools which have been described. This tariff, however, would, of necessity, fluctuate throughout under the pressure of competition at any one point. For instance, a lake-rate to be met at Chicago would affect the land-rate from Louisville; if it did not, one point would be discriminated against, as it now perpetually is, in favor of the other. In like manner a river-rate from St. Louis would affect the land-rate from Chicago. Thus the principle of the all-pervading action of competition would be generally established through a confederation, as it is locally established through combinations to-day. In this way, full effect would be given to that natural and healthy competition which is now so successfully localized, while railroad discrimination would be effectually repressed. Discrimination being thus disposed of, it would then only remain to guard against extortion. That would not, apparently, be difficult. In the first place, it would probably be found that the effect of natural competition would, once the play of its forces was made all-pervading, afford the necessary protection. If it did not, the extortion would have to be practiced openly, and by a responsible agent upon whom the whole force of public opinion might and would be directed. Should this fail to produce the desired effect, the central agency being responsible to the law as well as to public opinion, recourse could finally be had to legislation. Beyond this, it does not seem worth while at present to carry the discussion. The first step is, necessarily, to accustom the public mind to the idea that railroad combinations possibly may be an evil only because they are unrecognized, and that the proper way to deal with them may, perhaps, be through regulation and not through prohibition.

In pursuing the discussion, however, care must be taken lest the argument against competition is carried too far, or is not properly understood. It will not do to rush from one extreme to the other. The natural question which has already been suggested must be clearly borne in mind :—Why should a railroad combination, avowedly intended to hold competition in check, if not to put an end to it, produce any result other than the natural and obvious one of raising prices ?—Who or what is to protect the community against the extortions of these great corporations, should they cease to quarrel and compete among themselves?—And, in the first place, it must be frankly acknowledged that the argument against railroad competition can only be advanced subject to great limitations. Undoubtedly the fierce struggles between rival corporations which marked the history of railroad development, both here and in England, were very prominent factors in the work of forcing the systems of the two countries up to their present degree of efficiency. Railroad competition has been a great educator for railroad men. It has not only taught them how much they could do, but also how very cheaply they could do it. Under the strong stimulus of rivalry they have done not only what they declared were impossibilities, but what they really believed to be such. None the less, extraordinary as these results have been, they have been reached only at an excessive cost; a cost so excessive as to show clearly that the process is one which cannot be continued indefinitely. Under the incessant strain of competition the number of competitors is being steadily reduced. The present question, therefore, is not whether good results have ever been secured through railroad competition, but whether the same or even better results may not now be secured through other and less costly processes. During the last forty years the railroad system has grown, and experience has grown with it. During that time, also, competition has to a degree expended its force, and is now obviously working its way out to a final result. If that result is to be a legalized confederation it must be borne clearly in mind that, while railroad competition would cease, the influence of every other form of competition,—sea, lake, river and canal,—would through the machinery of that confederacy be economized and extended to its utmost possible limit. If the confederacy were touched by competition at one point, it would feel it at all points. Throughout, its rates would rise and fall together. Thus if one form of competition should cease, another would be prodigiously quickened.

But allowing even the monopoly to become complete, and having only such forms and degrees of restraint as law, usage, public sentiment and self-interest can supply, we are by no means without analogous cases having a very close bearing on the argument. In our cities, for instance, as regards the supply of gas, it is found cheaper and better for the community to have to do with one company than with several. So also as respects the supply of water. In this country it is now usual for cities and towns to construct their own water works. If this, however, were not the case, few would be disposed to deny that a city having to do with a single aqueduct company would be apt to have a much more satisfactory service than one which sought to divide it among many. Carrying now the argument directly into the case of railroads, and having recourse again to experience, we find that railroad competition has been tried all over the world, and that everywhere, consciously or unconsciously, but with one consent, it is slowly but surely being abandoned. In its place the principle of responsible and regulated monopoly is asserting itself. The same process, varied only by the differing economical, social, and political habits and modes of thought of the people, is going on in France, in Belgium, in Germany, and in Great Britain. The experience of the three first named countries bears much less strongly than that of England on the particular conditions existing in America, yet even for us their experience is not without its significance. In France we see six great corporations dividing the country into as many distinct territories, and each of the six directly responsible for the territory served by it ; while both these corporations and the government view with undisguised apprehension the recent appearance of a competing, though subsidiary, system. In Prussia, a plan of close government supervision through a special cabinet officer, a member of the ministry, is now being matured, while elsewhere in Germany the lines are rapidly passing into the hands of the governments, or under their more immediate control. Apart, however, from this, the German experience in one respect deserves peculiar notice. The local and governmental subdivisions of Germany, more than those of any other country of Europe, resemble our federated system of states. Placed in the centre of Europe, Germany is a species of thoroughfare, while at the same time the individual members of its railroad system belong under different jurisdictions. Here, then, every condition is found which is likely to incite an uncontrolled railroad competition. To a degree it existed, at one period, but the German temper and habits of thought are so different from the American that competition there speedily resulted in combination. The German railroad union, including as it does nearly one hundred different managements, operating twenty-six thousand miles of track, actually accomplishes many of the results which the Saratoga and Atlanta combinations were designed to accomplish. It makes all necessary arrangements respecting joint traffic, settles questions of fares and freight, and substitutes arbitration for wars of rates. It has to a certain degree introduced uniformity and stability into the system. The fact that such an association is easily formed in Germany, and is formed only with the greatest difficulty in America, proves nothing except the powerful influence of national thought and temper. A certain amount of waste and confusion sufficed to bring a system into being in the one case; the present question is, how much more waste and how much greater degree of confusion will be necessary to bring a somewhat similar system into being in the other case?

In Belgium alone has railroad competition proved a permanent advantage; and it has proved so there for the simple reason that the competition between railroads in Belgium, unlike that in the United States, was never uncontrolled. A hand was always on the regulator. The government, as the largest owner of railroads, was itself the chief competitor, and as such its action was certain, equitable, and justly distributed. It could not show preferences, or discriminate, or make good the losses sustained in fighting over a divided business out of profits exported from an exclusive business. Regulated in this way, competition could be kept alive and made beneficial. It did not wear itself out by its own excesses.

Of all foreign experiences, however, that of England most resembles our own. The only essential difference is that England is wealthier and infinitely more compact than the United States, so that, as respects railroads, causes produced their results much more quickly there than here. Nowhere, however, is the present tendency towards the concentration of railroad interests in a few hands more apparent than in England. The mill of competition has there about fulfilled its allotted work. The whole English railway system has now passed into the hands of a few great companies, by whom the country is practically divided into separate districts. These are literally in the hands of monopolies. The practical result of this consolidation, as compared with the old-fashioned competition, was set forth in two concrete cases by the parliamentary committee on railway amalgamation of 1872, in language which has already been quoted, but which in this connection will bear repetition.

The North-Eastern railway "is composed of thirty-seven lines, several of which formerly competed with each other. Before their amalgamation they had, generally speaking, high rates and fares, and low dividends. The system is now the most complete monopoly in the United Kingdom; from the Tyne to the Humber, with one local exception, it has the country to itself, and it has the lowest fares and the highest dividends of any large English railway. It has had little or no litigation with other companies. While complaints have been heard from Lancashire and Yorkshire, where there are so-called competing lines, no witness has appeared to complain of the North-Eastern; and the general feeling in the district it serves appears favorable to its management." *

There is scarcely a section of the United States which could not tell of an experience very like the English one just referred to. Massachusetts, for instance, could supply a well-known case in point. Of two sections of that state lying north and south of the city of Boston, the one known as the Cape Ann and the other as the Cape Cod district, the first has from the beginning been served by two rival lines whose whole history has been one long trial of strength, resulting at last in the absolute ruin of one and in the severe crippling of the other. How many millions of dollars were recklessly squandered in the long course of the struggle, it is impossible to compute. While the Cape Ann

---

* Report from Select Committee on Railway Companies Amalgamation (1872), page xxvii.

district has thus enjoyed the benefits of railroad competition, the southern or Cape Cod district has, on the other hand, been served by a single consolidated corporation, the cardinal principle with which has been monopoly. It appropriated to itself a certain district, and that district it undertook to furnish with all reasonable railroad facilities; but within the limits of its own territory it did not propose to tolerate any rival. The result in these two cases, whether in accordance with theory or not, is confirmatory of experience. Between its two rival corporations the northern district was through years converted into a battle-ground, and turned upside down; rates fluctuated wildly and varied everywhere; common tariffs were made and not observed, and profits were pooled; bits of connecting road were seized hold of by the one combatant or the other, and were perverted from serving the community into being engines of attack or defence. As to the two companies, with that impenetrable stupidity which usually characterizes the lover of petty independence, they sturdily preferred to lose thousands in conflict rather than incur the risk of being overreached in negotiation by so much as a dollar. Each of them absolutely threw away enough money to buy up the other in that stupid fighting in which thick-headed presidents and "smart" superintendents uniformly delight. The one meant "to get even with the other," and both were resolved, no matter how much it cost, to have its "share of the business." Between them they ruined the business, dissatisfied every one, and then—came to terms with each other. Meanwhile, in the south-eastern section of the state peace certainly prevailed, if not absolute contentment. As respects railroads this last it is not well to expect, and, if expected, it will not be found. Nevertheless it is certainly true that, according to general experience, the nearest approach to it is reached, not only abroad but here, through the course pursued in this case. The reliance on competition seems to give throughout a false direction to public opinion as respects railroads. They are looked upon as something alien, if not hostile. The public welfare is associated in the popular mind with their misfortunes. On the other hand the intelligent and peaceful operation of a consolidated company is generally followed by a sense of responsibility on the one side, and of ultimate friendliness on the other.

Besides the economical arguments which are so difficult to be overcome in this discussion, there are certain other objections to any such solution of the railroad problem as that suggested, which cannot be ignored. They have at least a strong hold on the popular ear and mind. In their character they are political or sentimental. As respects those of the first description, it is certainly not too much to say that jealousy of great corporations is a cardinal article in American political faith. There is reason for it, too; and in this respect recent scandals have given to railroad corporations a peculiar and unpleasant prominence. Neither is this instinctive jealousy confined to America. It is only a very few years since the present Sir Henry Tyler in one of the reports of the Board of Trade of Great Britain, formulated the proposition that the time was at hand when "the state must control the railroads or else the railroads would control the state." Yet when the parliamentary committee on amalgamations considered this question in 1872, they were obliged to report that the "growth of the corporations had not brought with it the evils generally anticipated." The fact is that in this, as in so many other instances, the truth of Mr. Disraeli's aphorism, that "in politics it is the unexpected which is apt to occur," received fresh illustration. In this country, as well as in Great Britain, those wise people who so earnestly point out the dangers incident to railroad concentration wholly ignore the important practical fact that concentration not only brings with it a corresponding increase of jealousy, but also an equally increased sense of responsibility. It is not the few great corporations which are politically dangerous, but the many log-rolling little ones. No one who has had experience in dealing before a legislative body with questions affecting railroad interests has failed to realize this fact. The burden of responsibility—almost of popular odium—which the large corporation bears, the ease with which a senseless cry can be raised against it, is even, as compared with smaller corporations, out of all proportion to its increased strength. So much has been written and declaimed on this subject, however, that it is well to be as distinct as possible in dealing with it. The popular apprehension of imaginary dangers to be apprehended from railroad consolidation is not well considered. With those who have most reflected on the subject it is safe to say that the idea of a combination of all the railroad interests of the country into the hands of three or four corporations,—even though they might practically be the creatures of a triumvirate's will,—would excite no apprehension. That corporation, or those who composed that triumvirate, would retain power only by most carefully abstaining from all abuse of power. Little as those who expatiate on the subject seem to realize it, it is nevertheless true that with each new railroad the Vanderbilt or the Jay Gould or the Huntingdon interest acquires, the more cautious and conservative they become. They realize the responsibilities and dangers of their position, if their critics do not. The only present difficulty is that those who undertake to represent the community neither understand the situation, nor know how to take advantage of it.

Finally it remains to consider the sentimental objections. The combination of railroads, it is claimed, is unrepublican,—through it the dynasty of the "Railroad Kings" is insidiously asserting itself. This argument is of the kind which sets refutation at defiance. Not infrequently it is met with in the columns of the press, but it is an argument appropriately addressed only to that discouragingly large class among whom words are money and not counters. It is unmitigated cant, and deserves only to be treated as such. There is a principle much nearer the foundation of republican institutions than any jealousy or apprehension of Railroad Kings— the great principle of not unnecessarily meddling. After all, men and systems can best develop themselves in their own way, and it is hardly worth while either to continually prognosticate evil, or to pass one's life in fighting shadows.

Briefly reviewing the whole ground which has now been traversed, it is obvious that the tendency of events and drift of discussion are everywhere the same—away from a reliance on the beneficial effect to be derived from the uncontrolled competition between railroads. In America only does any considerable body of reflecting persons continue to have faith in it. In France and in Belgium the principle never was recognized, and the later tendency is distinct and strong against its admission. In Great Britain, where it originated, it is now definitely abandoned. In Germany the highest authorities incline towards the idea of a confederation of railroads directly confronting the imperial government and responsible to it. The movement, however, seems to have its limits, and those tolerably well defined. The idea of state-ownership can hardly be said to be growing. On the contrary, in those countries like France and Germany, where recourse would naturally be had to this solution of the problem, the tendency seems now rather to be towards a close regulation of the railroads by the government, without its owning them: while in other countries, where the institutions are of a more popular character, a system of public supervision is assuming shape. Thus supervision would seem to be the limit of development on the one side, and regulation on the other.

Owing to the extremely complicated character of the American railroad system, rendering anything like a territorial division among corporations impossible, results here work their way out slowly. When they do work their way out, however, it is apt to be on a large scale and in a way not easily susceptible of change. So far as any progress has yet been made, it is obviously in the direction indicated,—the development of government supervision on the one side, and the concentration of railroads to escape competition on the other. The manner, indeed, in which, starting from different stand-points of interest and opposite sections of the country, the Massachusetts commission and the Southern Railroad and Steamship Association have unconsciously worked towards a common ground, is noticeable. On the one hand the whole effort of the commission has been to develop a tribunal which, in all questions affecting the relations of the railroad system to the community, should secure publicity and that correct understanding of the principles upon which only legislation of any permanent value can be based, and which is reached through intelligent public investigation. That secured, all else might safely be left to take its own course. A sufficient responsibility would be secured to afford a guarantee against abuse. On the other hand the fundamental idea of the association, without the realization of which it remains incomplete, is to so confederate the railroad system that the members of it should be amenable to control and that responsibility should attach to it. Could the two results be brought about, the machinery would be complete. The confederated railroad system would confront the government tribunal, and be directly responsible to public opinion. This is almost precisely the result arrived at in France and in Great Britain, and is that contemplated in Germany.

It would be altogether premature to predict with any confidence that this or a similar result will speedily be reached in this country. Judging by experience, it is more probable that the development on the side of the railroad system will far outstrip that on the side of the government. The popular disbelief in the possibility of any permanent combination of the railroads, at once general and effective, is so complete that no provision will be made for it. Should one be brought about it will, however, in all probability, once it assumes shape, assume it very rapidly. In that case no great degree of public injury would necessarily be sustained, but the difficulty of thereafter restoring the necessary equilibrium would be materially increased. Another and more persistent political movement of the Granger character might and probably would become a necessity. As opposed, however, to an overshadowing commercial interest, so concentrated that all eyes and passions could be brought to bear upon it, this is not likely to be a movement difficult to originate or easy to resist.

END.

Economic Monographs:

A Series of Essays by representative writers, on subjects connected with Trade, Finance, and Political Economy.

I. **WHY WE TRADE, AND HOW WE TRADE;** or an Enquiry into the Extent to which the existing Commercial and Fiscal Policy of the United States Restricts the Material Prosperity and Development of the Country. By DAVID A. WELLS. 8vo, paper, ... 25 cents.

II. **THE SILVER QUESTION.** The Dollar of the Fathers *versus* the Dollar of the Sons. By DAVID A. WELLS. 8vo, paper, ... 25 cents.

III. **THE TARIFF QUESTION** and its Relation to the Present Commercial Crisis. By HORACE WHITE. 8vo, paper, ... 25 cents.

IV. **FRIENDLY SERMONS TO PROTECTIONIST MANUFACTURERS.** By J. S. MOORE. 8vo, paper, ... 25 cents.

V. **OUR REVENUE SYSTEM AND THE CIVIL SERVICE: Shall They be Reformed?** By ABRAHAM L. EARLE. 8vo, paper, ... 25 cents.

VI. **FREE SHIPS: The Restoration of the American Carrying Trade.** By Captain JOHN CODMAN. 8vo, paper, ... 25 cents.

VII. **SUFFRAGE IN CITIES.** By SIMON STERNE. 8vo, ... 25 cents.

VIII. **PROTECTION AND REVENUE IN 1877.** By Prof. W. G. SUMNER, Author of "History of Protection in the United States. 8vo, paper, ... 25 cents.

IX. **FRANCE AND THE UNITED STATES:** Their Commercial Relations considered, with Reference to a proposed Treaty of Reciprocity. A Series of Papers by PARKE GODWIN, M. MENIER, LEON CHOTTEAU, and J. S. MOORE. 8vo, paper, ... 25 cents

PUBLICATIONS OF G. P. PUTNAM'S SONS.

FOR LIBRARIES, TEACHERS, STUDENTS, AND FAMILY USE.
COMPREHENSIVE, COMPACT AND CONVENIENT
FOR REFERENCE.

THE HOME ENCYCLOPÆDIA
OF BIOGRAPHY, HISTORY, LITERATURE, CHRONOLOGY AND ESSENTIAL FACTS.
COMPRISED IN TWO PARTS.

Price in Cloth, $9 50; in half Morocco, $14 50.
SOLD SEPARATELY OR TOGETHER.

PART I
THE WORLD'S PROGRESS

A Dictionary of Dates, being a Chronological and Alphabetical Record of all Essential facts in the Progress of Society, from the beginning of History to August, 1877. With Chronological Tables, Biographical Index, and a Chart of History,

By G. P. PUTNAM, A.M.

Revised and continued by F. B. PERKINS. In one handsome octavo volume of 1,000 pages, cloth extra, $4.50; half morocco, $7.00.

CONTENTS:

- THE WORLD'S PROGRESS, 1867—1877.
- THE SAME 1851—1867.
- THE SAME FROM THE BEGINNING OF HISTORY TO 1851.
- UNITED STATES TREASURY STATISTICS.
- LITERARY CHRONOLOGY, ARRANGED IN TABLES: HEBREW, GREEK, LATIN AND ITALIAN, BRITISH, GERMAN, FRENCH, SPANISH AND PORTUGUESE, DUTCH, SWEDISH, DANISH, POLISH, RUSSIAN,
- ARABIAN, PERSIAN AND TURKISH, AMERICAN.
- HEATHEN DEITIES AND HEROES AND HEROINES OF ANTIQUITY.
- TABULAR VIEWS OF UNIVERSAL HISTORY.
- BIOGRAPHICAL INDEX, GENERAL.
- THE SAME OF ARTISTS.
- SCHOOLS OF PAINTING IN CHRONOLOGICAL TABLES.

"A more convenient labor-saving machine than this excellent compilation can scarcely be found in any language."—*N. Y. Tribune.*

"The largest amount of information in the smallest possible compass."—*Buffalo Courier.*

"The best manual of the kind in the English language."—*Boston Courier.*

"Well-nigh indispensable to a very large portion of the community."—*N.Y. Courier & Enquirer.*

PART II
THE CYCLOPÆDIA OF BIOGRAPHY
A RECORD OF THE LIVES OF EMINENT MEN
By PARKE GODWIN.

New edition, revised and continued to August, 1877. Octavo, containing 1200 pages, cloth, $5.00; half morocco, $7.50.

The Publishers claim for this work that it presents an admirable combination of compactness and comprehensiveness. The previous editions have recommended themselves to the public favor, as well for the fulness of their lists of essential names, as for the accuracy of the material given. The present edition will, it is believed, be found still more satisfactory as to these points, and possesses for American readers the special advantage over similar English works, in the full proportion of space given to eminent American names.

PUBLICATIONS OF G. P. PUTNAM'S SONS.

Standard Works of Reference.

PUTNAM (GEORGE PALMER) **The World's Progress.** A Dictionary of Dates. Being a Chronological and Alphabetical Record of the essential facts in the progress of Society. With Tabular views of Universal History, Literary Chronology, Biographical Index, etc., etc. From the Creation of the World to August, 1877. By GEORGE P PUTNAM. Revised and continued by FREDERIC BEECHER PERKINS. Octavo, containing about 1,200 pages, half morocco, $7 00; cloth extra, ... $4 50

** The most comprehensive book of its size and price in the language.
"It is absolutey essential to the desk of every merchant, and the table of every student and professional man."—*Christian Inquirer.*
"It is worth ten times its price. * * * It completely supplies my need."—S. W. PIEGART, *Principal of High School, Lancaster, Pa.*
"A more convenient literary labor-saving machine than this excellent compilation can scarcely be found in any language."—*N. Y. Tribune.*

HAYDN. A Dictionary of Dates, relating to all Ages and Nations, for Universal Reference. By BENJAMIN VINCENT. The new (15th) English edition. With an American Supplement, containing about 200 additional pages, including American Topics and a copious Biographical Index, by G. P. PUTNAM, A. M. Large Octavo, 1,000 pages. Cloth $9 00; half russia . . . $12 00

THE BEST READING. A classified bibliography for easy reference. Edited by FREDERIC B. PERKINS. Fifteenth edition, revised, enlarged and entirely re-written. Continued to August, 1876. Octavo, cloth, $1 75; paper $1 25

"The best work of the kind we have ever seen."—*College Courant.*
"We know of no manual that can take its place as a guide to the selection of a library."—*N. Y. Independent.*

PUTNAM'S LIBRARY COMPANION A quarterly summary, giving priced and classified lists of the English and American publications of the past three months, with the addition of brief analyses or characterizations of the more important works; being a quarterly continuation of THE BEST READING. Published in April, July, October, and January. Price to subscribers, 50 cts., a year. Vol. I., boards, 50 cts.

"We welcome the first number of this little quarterly. It should prove invaluable alike to librarians, to students, and to general readers."—*Boston Traveler.*

JUKES (THE) A STUDY IN CRIME, PAUPERISM, DISEASE, AND HEREDITY. By R. L. DUGDALE. Published for the "Prison Association of New York." Octavo, cloth $1 25

"A work that will command the interest of the philanthropist and the social reformer, and deserves the attention of every citizen and taxpayer."—*N. Y. Tribune.*

JERVIS (JOHN B.) **Labor and Capital.** A complete and comprehensive treatise by the veteran engineer, whose experience of more than half a century has given him exceptional opportunities for arriving at a practical understanding of the questions now at issue between employers and employed. 12mo, cloth $1 25

LINDERMAN (HENRY R., Director of the United States Mint) **Money and Legal Tender in the United States.** 12mo, cloth 1 25

PUBLICATIONS OF G. P. PUTNAM'S SONS.

CONSTANTINOPLE. By EDMUNDO DE AMICIS, author of "A Journey through Holland," "Spain and the Spaniards," &c. Translated by CAROLINE TILTON. With introduction by Prof. VINCENZO BOTTA. Octavo, cloth.

A trustworthy and exceptionally vivid description of the city which, in the present reopening of the Eastern question, is attracting more attention than any other in the world. De Amicis is one of the strongest and most brilliant of the present generation of Italian writers, and this latest work from his pen, as well from the picturesqueness of its descriptions as for its skilful analysis of the traits and characteristics of the medley of races represented in the Turkish capital, possesses an exceptional interest and value.

THE GREEKS OF TO-DAY. By Hon. CHARLES K. TUCKERMAN, late Minister Resident of the U. S. at Athens. Third Edition. 12mo, cloth, ... $1.50

This work attracted special attention at the time of its publication, in 1872, as giving a trustworthy and interesting picture of life in Greece, and of the character and status of the modern Greek. At this time, when public attention is so generally directed towards the scheme of practically re-establishing a Greek empire and Greek supremacy in the East, it is thought that a new edition will prove of interest and service.

"The information contained in the volume is ample and various, and it cannot fail to hold a high rank among the authorities on modern Greece."—*N. Y. Tribune.*

"No one can read this book without having his interest greatly increased in this brave, brilliant, and in every way remarkable people."—*N. Y. Times.*

"We know of no book which so combines freshness and fullness of information."—*N. Y. Nation.*

ENGLAND; POLITICAL AND SOCIAL. By AUGUSTE LAUGEL. Translated by J. M. HART. 12mo, cloth, ... $1.50

"It is written with a tone of confidence and force of expression which captivate."—*Buffalo Commercial.*

"Affords a clear, distinct, and comprehensive view of the political institutions of England."—*N. Y. Nation.*

"Here, in every sense, is a charming book. * * * So full of thought, that, like the best of Macaulay's Essays, it will bear reading more than once. * * * We have rarely met with more picture-like descriptions of what seems to have dwelt most upon his mind—English landscape scenery and rural life."—*N. Y. World.*

THE SILVER COUNTRY; or, THE GREAT SOUTHWEST. A Review of the Mineral and other Wealth, with the attractions and material development of the former kingdom of New Spain, comprising Mexico and the territory ceded by Mexico to the United States in 1848 and 1853. By ALEXANDER D. ANDERSON. 8vo, cloth, with Hypsometric Map, ... $1.75

"Just at the present moment everything which affords reliable information on the question of silver, its uses and production, is of almost paramount interest."—*Washington National Republican.*

"A very useful book for those who wish to study the silver question in its fundamental feature."—*Chicago Journal.*

"The book will unquestionably become the authority on the subject of which it treats."—*St. Louis Republican.*

www.ingramcontent.com/pod-product-compliance
Lightning Source LLC
Chambersburg PA
CBHW031816230426
43669CB00009B/1166